AFTERMATH

ALSO BY LAWRENCE E. JOSEPH

Gaia: The Growth of an Idea (1990)

Common Sense: Why It's No Longer Common (1994)

Apocalypse 2012: An Investigation into Civilization's End (2007)

LAWRENCE E. JOSEPH

Broadway Books / New York

A GUIDE TO PREPARING

FOR AND SURVIVING

APOCALYPSE 2012

AFTERMATH

Library of Congress Cataloging-in-Publication Data

Joseph, Lawrence E.
    Aftermath / Lawrence E. Joseph.—1st ed.
        p.   cm.
    1. Two thousand twelve, A.D.   2. Twenty-first century—Forecasts.
3. Catastrophical, The.   4. Regression (Civilization)   I. Title.
    CB161.J67 2010
    303.49—dc22

                                                        2009043227

ISBN 978-0-7679-3078-9

Printed in the United States of America

DESIGN BY ELINA D. NUDELMAN

10 9 8 7 6 5 4 3 2 1

First Edition

To Phoebe and Milo, I love you even more.

To my mother, Yvonne Joseph, who tells me the truth nicely. Usually.

To the Leaning Tower of Pisa, which, though built on shifting sands, has defied the pull of gravity for over eight hundred years and now, through the miracle of modern engineering, seems stabilized for several hundred more. We should all be so fortunate.

CONTENTS

AFTERMATH

So here's the situation: we have received a tip that the year 2012 is going to be a lulu, quite possibly catastrophic to an unprecedented degree. The source is an ancient Mayan prophecy that the date 12/21/12 will signal the birth of a new age, with all the joy, blood, and pain that would naturally accompany any birth. Few of us know much about Mayan culture, and under normal circumstances we could probably just ignore the whole thing, were it not for three important facts: (1) the Maya have been right before, (2) the Sun is mounting toward a climax of unprecedented ferocity in 2012 at the same time, perhaps coincidentally, that the Earth's protective magnetic shield against such solar outbursts is weakening dramatically, and (3) apocalypse is in the air.

THE MAYA WERE RIGHT BEFORE

After five years of research, first for Apocalypse 2012: An Investigation into Civilization's End (2007) and then for this book, I have come to the opinion that the ancient Mayan prophecies regarding 2012 should not be accepted as divine decree but neither should they be dismissed because they are not conventionally scientific. The year is best understood as the crescendo of a metamorphic process already well under way. It is not necessarily a firm deadline but rather a compelling shorthand for the possibility of civilization-altering cataclysm and/or revelation occurring within our lifetimes.

AFTERMATH

At bottom, the Mayan prophecy for 2012 is based on a simple astronomical observation. On 12/21/12, the day of the winter solstice, the Sun will eclipse, that is to say, interpose itself, between the dead center of the Milky Way galaxy and the Earth. Imagine a perfect cross, or crosshairs, forming between the Sun's meridian (vertical equator) and the Milky Way's ecliptic (horizontal equator) on that perhaps fateful date. This celestial formation, which the Maya call "The Tree of Life," planted right in the black womb of the galaxy, occurs once every 5,200 years, a period called a "sun" or an "age." On 12/21/12, we will be entering the Fifth Age, called "Job Ajaw," an age destined to restore balance and bring enlightenment to humanity, eventually.

For the Maya, 12/21/12 is the celestial equivalent of the clock striking midnight, for the beginning of a new age. The rest of it—the joy, blood, and pain part, and all the predictions of catastrophe and revelation that have gone along with it—is all interpretation and lore. It could be that 12/21/12 proves as inconsequential as any other New Year's Eve, an arbitrary if useful demarcation of time and therefore a good excuse for a party, but nothing more than that, no real physical change of any significance.

In order to evaluate the validity of the Maya's 2012 prophecy, it is necessary to examine the track record of other determinations made using the same astronomical system. More than a thousand years ago, during what is called the Classic Mayan culture, Mayan astronomers calculated the length of the solar year to within eight-tenths of a second of what we, with all our supercomputerized technology, now determine it to be. In a nutshell, the Maya understood the cosmos as a five-dimensional space-time continuum called "najt," composed of the conventional three dimensions, a fourth dimension of time (as Einstein proposed), and a fifth dimension, frequency, or level of vibration. Using this theoretical framework, they proposed the existence of "wormholes," vibrational tunnels through the space-time continuum that contemporary astrophysicists have also hypothesized and whose physical properties they have just now begun to investigate. The importance for us is that events occurring unfathomably far away can and do have a real-world impact on our planet and its inhabitants.

"We inhabit a reality shaped and modeled after the cosmos, and we vibrate in tune with the movement of the stars," writes Carlos Barrios in *The Book of Destiny* (HarperOne, 2009). Barrios is a Mayan shaman from Antigua, Guatemala, whom I have known and worked with for several years. This, of course, is the basic principle that underlies astrology, practiced by all civilizations. Hard to argue with such lofty sentiments, but how, one wants to know, does Mayan astrology translate into this-world relevance and accuracy? Centuries in advance the Maya accurately predicted that on a date known as One Reed, Ce Acatal, or, in our Gregorian calendar terms, Easter Sunday, April 21, 1519, their civilization would be devastated by butterflies. These butterflies turned out to be the sails of Hernando Cortés's eleven Spanish galleons as they rounded the horizon. Thus began the Spanish conquest of what is now Latin America. A bloody good guess.

Even more striking is the ancient Mayan prediction made almost two millennia ago that on 12/21/12 humanity would move into the Ethereal Age. With the launch of the Sputnik satellite in 1957, that's exactly what humanity began to do, move into the ether, the stuff of outer space, the void that, though largely a vacuum, is anything but meaningless or useless. Okay, so the prophecy made two thousand years ago was fifty-five years off. The timing and essence of the foresight derived from their astronomical observations was preternaturally apt. Today, just a few years shy of 2012, virtually all of our telecommunications, military security, law enforcement, finance, and trade is conducted by satellites orbiting in the ether of outer space. It is no stretch to say that humanity has now entered the Ethereal Age.

THE SOLAR CLIMAX OF 2012

From time immemorial, the Sun has pummeled the Earth with giant blasts of radiation, and usually we are none the worse for wear. Upon colliding with the Earth's protective magnetic field, these blasts, known as coronal mass ejections (CMEs), create beautiful auroras, also known as the northern lights. It may also be that the biosphere makes use of these radiation inoculations in some constructive way. CMEs peak and ebb in frequency and intensity over an eleven-year cycle, known as the solar cycle. Occasionally these peaks are extraordinarily intense, as has occurred several times over the past century and a half. In 1859, the northern lights were visible all the way down to the equator, so brilliant that one could read a book by them at midnight over much of the Northern Hemisphere. This event was far more spectacular than destructive, aside from a few fires and a rash of disruptions in telegraph service. But back then, society did not run on electricity, was not stitched together by power grids that would have shorted out all around the globe.

Were a solar event of that magnitude to happen today, global civilization would be devastated. According to a recent report, *Severe Space Weather Events*, issued in December 2008 by the National Research Council of the National Academy of Sciences: "Impacts would be felt on interdependent infrastructures with, for example, potable water distribution affected within several hours; perishable foods and medications lost in 12–24 hours; immediate or eventual loss of heating/air conditioning, sewage disposal, phone service, transportation, fuel resupply and so on. These outages would probably take months to fix, straining emergency services, banking and trade, and even command and control of the military and law enforcement."

The next climax of solar activity will, by scientific consensus, come in 2012. At present, there are two competing theories regarding the coming climax, one that it will be below average in intensity, and the other, which seems to have more support among solar physicists, that it will be above average, remarkably so. The only thing everyone agrees on is that 2012 will be abnormal.

One way or the other, it looks like we are in for a beating. The National Academy of Sciences' frightening report was issued before an astonishingly large hole in the Earth's protective magnetic field was discovered by THEMIS, a group of five NASA scientific research satellites that flew through a hitherto undiscovered pole-to-equator gap in December 2008. Our planet's magnetic field is our sole defense against solar blasts, so the sudden appearance of immense holes in that field is a cause for real concern. Moreover, conclusions drawn from the data stream provided by those satellites indicate that the magnetic fields of the Earth and the Sun will almost certainly align themselves in such a way to maximize the impact of solar blasts, just in time for the 2012 climax. It's just bad luck. As I'll explore in Section I of this book, the perfect space storm is headed our way.

APOCALYPSE IS IN THE AIR

"It's about goddamn time!" growled Kurt Vonnegut, when informed that a growing number of people believe that the Apocalypse may well come in 2012.

Public figures from across the ideological and cultural spectra have cited the possibility that 2012 could be the end of our civilization as we know it. United States Senator Bernie Sanders of Vermont, a left-winger who caucuses with the Democratic Party, argues that preparing for disaster in 2012 may well be the most constructive response to the threat, providing jobs now and vital defenses against catastrophes that, 2012-related or not, will likely come within our lifetimes as a result of runaway global warming, terrorist nuclear catastrophe, and/or global pandemics. At the other political extreme, Governor Mike Huckabee of Arkansas, who in 2008 ran a spirited right-wing campaign for the Republican nomination for president of the United States, has warned that 2012 could well see the global economy shut down for months. And Tim LaHaye, Christian fundamentalist minister and immensely influential author of the megabestselling *Left Behind* series of books about the coming Armageddon, has been boldly supportive of my recent book warning of disaster in that year.

End-of-the-world 2012 fantasies engulf the popular media as well. Hundreds and hundreds of books, blogs, video games, documentaries, and films, the splashiest of which is Roland Emmerich's 2012, a $200 million Sony Pictures extravaganza. "Never before has a date in history been so significant to so many cultures, so many religions, scientists and governments. '2012' is an epic adventure about a global cataclysm that brings an end to the world and tells of the heroic struggle of the survivors," goes the official plot summary. I suppose you could think of this book as a nonfiction version of Emmerich's film, except that it's stuffed with a lot more facts than action, and the heroic survivors are, I hope, us.

There's no denying that apocalypse is in the air these days, quite literally. Greenhouse gases are responsible in part or whole for global warming–related disasters ranging from superpowerful hurricanes to rampant desertification. Scenarios of future climate change cataclysms, such as the poles melting and sea levels rising to inundate major cities, have been drummed into our heads graphically and insistently. Climate-driven apocalypse seems on its way, the only question being when it will descend upon us.

Nothing blows the mind more than spectacular failure, and no failure has been more spectacular than that of our collective brain-trust in the run-up to the stock market crash of September 2008. Hell, Harvard University, the brainiest place in the world, was completely blindsided, lost a third of its $26 billion portfolio even though it was managed by at least a hundred Ivy-degreed, lavishly compensated bankers and advisors, no doubt with giant IQs and terabytes of all the latest high-tech information at their fingertips. Harvard got hit so bad it had to emergency-borrow a billion and a half at unfavorable terms just to keep afloat.

That the Treasury Department and the U.S. Securities and Exchange Commission both failed miserably in their job to protect the public is, of course, quite scandalous and sad, though government corruption and incompetence, even on such a massive level, is hardly astonishing. But when branches of the government fall rotten, it is only natural to wonder nervously whether the whole tree might be getting ready to crash down. Was the 2008 stock market crash the financial precursor of a broader systemic collapse in 2012? The four-year lead time would seem to fit with two other great economic collapses of the past hundred years. The great stock market crash of 1929 bottomed out about four years later in 1933–1934. The other great crash of the twentieth century, in 1987, bottomed out about four years later, in 1990–1991. If the economic collapse of 2008, perhaps as severe as 1929 and certainly worse than 1987, follows history, it will hit rock bottom late in 2012.

Which path will we follow? The Great Depression of the early 1930s was a global backslide into poverty that in turn led to the Second World War, the most hideous carnage humanity has ever endured. The recession of the early 1990s led to the dot.com boom, one of the greatest periods of creativity and prosperity in history. The financial crisis that began in 2008 could turn out either way, or go in some new direction entirely. Aristotle taught that all good tends toward the middle and therefore might have offered the hope that 2012 finds us having navigated a course between the haywire consumerism that contributed to the crash and the rampant social chaos that could ultimately result from our inability to keep on purchasing. Another depression or a third world war would of course be an unfathomably dismal outcome. But so, I believe, would be a prompt and neat recovery from the 2008 market crash if it meant relapsing into shopaholic consumption, a soul-deadening addiction.

THE CHOICE FOR 2012: ENLIGHTENMENT OR MATERIALISM?

If we don't curb our crazy consumption, if we don't stop trying to fulfill deep emotional and spiritual needs with credit cards, and worse, going into ruinous debt to feed that soulless compulsion, civilization will emerge from 2012 crippled beyond repair. According to Carlos Barrios, the middle course between greed and privation is there for the taking. Ever since 1987, year of the Harmonic Convergence of planets as well as the major stock market crash, we have been in a time when "the right arm of the materialistic world is disappearing," writes Barrios. According to the Mayan vision, 2012 is a chance to set things right. It is that once-in-every-26,000-years opportunity to pass through a door in time to a less materialistic and more healthful and compassionate way of living.

My compliments to those ready to walk joyously through that door. Over the past several years, I have received hundreds of messages from people all over the world who welcome the prospect of transcendence to this higher, more harmonious plane of being, some so enthusiastically that they are willing, even eager, to suffer whatever pain and privation such a glorious transition might require. Would you be willing to suffer poverty or injury to live on a more godly plane? Not me. I loved *Inferno* but just could never get into *Paradise*. I would take Friday afternoon soccer practice over a chance for spiritual enlightenment any day. I like my cushy life, my house, my imported Bulgarian feta cheese. My two young kids are growing up safe and happy, and expect to go to Sea World every year to watch the killer whales do their pirouettes.

There's no denying that a robust economy acts as sort of an insurance policy for society, providing the budget to prevent and/or respond to natural and man-made disasters. Rampant materialism is what pays the premiums. If we are blindsided while struggling to revive a bankrupt economy—if a comet hits, if a supervolcano blows, if weapons of mass destruction are used genocidally, if a war between major powers breaks out, if solar flares or military attacks bring down our satellite system, if famine or plague become pandemic—the dominoes will fall all over the world. Natural catastrophe such as the Great Flood leads to political instability leads to chaos, which religious fanatics take as a sign to perpetrate holy hell. Solar storms cripple the satellite system, which pokes holes in military and law enforcement surveillance, enabling terrorists and/or rogue states to attack. Famine caused by climate change and spiking food prices weakens the immune systems of the hungry multitudes, leading to plague, which would in turn lead to deeper famine, since local agricultural labor forces would be seriously weakened by the spreading disease. And wouldn't all that chaos be just the perfect flashpoint for Armageddon, the war-to-end-all-wars Bible prophecy that so many people in the Middle East seem intent on fulfilling? Who in his right mind could opt for that?

However, as Barrios explains the Mayan cosmology, the real choice for 2012 is not between enlightenment and materialism but between enlightenment and, well, being a materialistic, warlike loser clinging to obsolete ways. Get with the 2012 program or get left behind altogether, is the Mayan bottom line.

"THROW THE EGGHEADS OUT!"

Periodically it happens pretty much across the board to politicians, so why not to intellectuals? The acutely embarrassing failure of virtually everyone with any intellectual standing in our society to foresee the colossal collapse of the global economy in 2008 has had spine-chilling ramifications far beyond the realm of finance and governance, calling into question the fundamental competence of our current crop of marquee intellectuals. There's a budding anti-incumbency resentment against "high-IQ idiots," those designated by their position in government, academe, or the serious news media to apprise us of the basic threats and opportunities facing the human community.

A new wave of scholarship on global catastrophe has flourished unnervingly. Alan Weisman's bestselling book, The World Without Us, leaves us with a sense that civilization is fragile, that the blessings of daily life—electricity, clean water, the ability to communicate beyond personal distance—are not to be taken for granted. James Lovelock, in his latest book, Gaia's Revenge, observes that about thirty civilizations have come and gone over the past five thousand years, implying, of course, that ours could be next. In Our Final Hour, Martin Rees, Britain's Astronomer Royal, gives a grim and erudite rundown of all the ways daily life could, and likely will, be done in. Jared Diamond's Collapse provides parables of past civilizations gone wrong, drawing numerous parallels to our own. Global Catastrophes and Trends, by Václav Smil, and Global Catastrophic Risks, edited by Nick Bostrom and Milan M. Cirkovic, essentially the proceedings of a June 2008 Oxford University symposium, explore the realm of low-frequency/high-consequence (LF/HC) events, each taking the "when, not if" approach to anticipating cataclysm.

"People who would never dream of hurting a child hear of an existential risk [megadeath scenario] and say, 'Well, maybe the human species doesn't really deserve to survive,'" writes Eliezer S. Yudkowsky in *Global Catastrophic Risks*.

What an intellectual about-face from the past few decades of rosy futurism! Take Herman Kahn, the megamind behind the Hudson Institute who is considered the father of futurism. In 1967, Kahn and coauthor Anthony Wiener published *The Year 2000*, in which they opined that global civilization had become so advanced and sophisticated that we had entered an era in which the future from there on out would essentially progress "surprise-free"... Surprise!

In *Future Shock*, Alvin Toffler accurately foretold of the rising beneficent influence of the personal computer in our lives, no mean feat. But Toffler missed, entirely, the energy issue, this despite the fact that his breakthrough bestseller was published just before the Arab oil embargo of 1973, a true future shock that sent gasoline prices soaring and inaugurated the latest chapter of the global struggle for energy resources that has shaped so much of our geopolitical reality ever since. Even more confident, neoconservative Francis Fukuyama declared in 1992 that history as we know it is over except for the shouting, because there can be no outcome other than the triumph of liberal democracy. Then came militant Islam...

Why the sudden shift from confetti to crash helmets? The dirty little secret of global interdependence is that its flip side is global vulnerability. What befalls one part of the world befalls the rest, faster and harder than ever before. One needn't be a hard-core pessimist, therefore, to wonder if maybe we're going to have to face-plant a few more times before we get this globalization thing down pat. Heck, it's only been a century since World War I, the planet-wide slugfest whereby the far corners of the world, from Berlin to Samoa, Cape Town to Tokyo, bareknuckle introduced themselves to each other. On the subject of world wars, it's hard not to wonder about the ancient biblical superstition that everything happens in threes.

The New Catastrophism, as I think of this emerging school of thought, signals a shift among scholars away from their tendency to predict more of the same, based on their analysis of history and current trends. Their need to justify their conclusions as rigorously as possible, lest they suffer debilitating criticism from their peers, has traditionally made them timid in forecasting radical departures, which, by nature, are less amenable to substantiation by historical fact. Recently, however, several intellectual countercurrents have helped open up the academic mind to more radical possibilities. Chaos and catastrophe theorists have tried nobly to analyze what gives rise to radical changes and breaks with the past. Edward Lorenz's explication of the "butterfly effect," whereby tiny inputs strategically positioned in space and time can trigger mammoth changes, is now, thanks to James Gleick's bestselling book *Chaos: Making a New Science*, part of the intellectual vocabulary. Chaos theory is cousin to catastrophe theory in that both examine the ways in which systems collapse or change abruptly. Developed by René Thom, the late French mathematician who won the Fields Medal for his work on the subject, and E. C. Zeeman, the British philosopher who has translated much of Thom's work for the lay intelligentsia, catastrophe theory anatomizes transformative events into categories. For example, "cusp" catastrophes are those which can go in one of two directions when a threshold is reached, such as when a dog is sufficiently irritated to either attack or be cowed. A "fold" catastrophe is when a system folds in on itself, the simplest example being the length at which a given grade of paper, held between the fingertips at one edge, will no longer stick out taut, but instead droop and fold. The chaos/catastrophe approach has now become something of an intellectual vogue, with many more scholars than ever before striving to explain the (seemingly) unexplainable and predict the (damn near) unpredictable. Most notable recently is Nassim Nicholas Taleb, whose "black swan" phrase for the out-of-the-blue appearance of something significant is fast becoming part of the scholarly lexicon.

These days, the sky seems full of black swans, flying in formation. There is a growing sense, perhaps overblown, that we are headed for a dominolike collapse of civilization triggered by a pile-up of threats ranging from asteroid and comet impacts (far more frequent and devastating than previously believed), to mutant nanotechnology (microscopic robots running amuck), to religious Armageddon (the goal of Middle Eastern extremists of all stripes), to supervolcano eruption (the last one, Lake Toba in Indonesia, killed about 90 percent of human beings on the planet), to famine and plague (a vicious cycle that once plunged the Middle Ages into a century of barbarism), to satellite system shutdown (the easiest way, militarily, to bring civilization to its knees), to bizarre solar behavior (which could well cause the collapse of the electrical power grid), to the metastasizing proliferation of weapons of mass destruction.

THE DOOM-SAYING IMPULSE

The good news is that Isaac Newton predicted that, at the earliest, the world would end in AD 2060, before which time at least half of us currently alive will be safely in our graves. One of the most brilliant minds in the history of our species, discoverer of gravity and of the three basic physical laws that govern all but the macro and micro extremes of the universe, inventor of the reflecting telescope, codeveloper of calculus,

Newton felt compelled, after years of studying the Bible, teaching himself Hebrew and Greek just so he could get it exactly right and keeping 4,500 pages of notes, to allow how we might even make it all the way to 2344. That's long enough for everyone alive today, even babies born right this second, to live to see their great-great-great-great-grandchildren, should they be so blessed.

Question is, what's the countdown like?

May the year 2012 turn out to be like 1936 and 1975, both dates when Kingdom was prophesied to come but didn't. Herbert W. Armstrong, founder of the Worldwide Church of God, often credited with starting the Christian evangelical/fundamentalist revival in the United States, taught that only loyal members of his church would experience the Rapture, in which devout and doctrinaire Christians expect to be drawn up to heaven to see Jesus while the rest of us are left on Earth to the horrors of Armageddon and Apocalypse. Armstrong first predicted that the Rapture would occur in 1936, but when Kingdom did not come, Armstrong recalculated and came up with a new deadline of 1975, which also passed more or less peaceably. That Armstrong was quite wrong in his predictions did little to stop his Worldwide Church of God from becoming an immensely wealthy and influential global enterprise.

"All will die except those who heed what I have to say," was Armstrong's metamessage. The specific date was less important than the remedy, which was to believe, believe, believe.

What compels human beings to prophesy the end of the world? Doomsayers such as Armstrong frequently suffer from inflated self-importance, considering themselves uniquely worthy, chosen by God or fate to possess the most important information that could possibly exist—the secret of the end of the world. This grandiosity is particularly true of those who prophesy end-dates expected to fall within their own lifetimes and/or those of their adherents, because such predictions by their very nature demand radical action, prescribed, of course, by the doomsayer. No doubt that many doomsayers predict the end of the world just to get noticed, while others are sincere in their belief. For my part, I am convinced that the 2012 prophecies are too serious to be ignored.

Albert Einstein is said to have remarked that he preferred expressing himself in mathematics over natural language such as English or German, because with mathematics he could create whatever language was necessary to express his thoughts while with natural language he was confined to the set of preexisting words. Perhaps some doom prophecies derive from a similar frustration with linguistic confines, kind of the way kids sometimes bump up against the ceiling of superlatives when trying to express the enormity of something, and end up using terms such as "googleplex infinity," "jillion-gazillion." Not to imply that doomsday predictions are necessarily childish, just that they are the rhetorical equivalent of shouting at the top of one's lungs.

So why do people listen? Perhaps the notion that the world will come to an end in 2012 or, more generally, one day soon, is nothing but the latest flare-up of our species' chronic need to scare itself silly, just as individuals do with horror films, death-defying sports, and other dopamine-inducing activities. Looking death in the eye helps us vent our frustration and inadequacy, to not be bored by the endless procession of days. Doom prophecies are how we collectively go mano-a-mano with mortality, how we grasp what Frank Kermode, the Oxford literary scholar, calls "the sense of an ending," not just of ourselves but of life as we know it, of humankind.

Each of us is to some degree prone to the misconception that in this life there exist two equally important and significant entities: (1) oneself, and (2) everyone/everything else. Martin Buber called it the "I-Thou" relationship, which, at its best, represents the loving bond between an individual and the rest of existence. Our fear for the welfare of the "Thou" therefore projects itself crazily at times. Predictions of doom may result from overempathizing with the "Thou" part by attributing the time frame of our own mortality, normally a century or less, to humankind, whose lifespan is many hundreds of centuries, or even to the Earth itself, whose lifespan is many millions of centuries. Our intellectual inability to grasp such immense time spans tends to confer a false sense of finitude to life.

The question of destiny is nonetheless on many minds as we approach 2012. The notion that the world might come to an end sometime soon inevitably focuses attention on the quality of the moments that remain. More important than the accuracy of the apocalypse prediction is the spirit in which it is made. Does the prophecy focus the Zeitgeist constructively or does it simply fan the flames of fear? What actions do the prophet's words motivate: drinking poisoned Kool-Aid or volunteering for the Red Cross?

A SELF-FULFILLING PROPHECY?

Warning about 2012 and its aftermath could, of course, backfire big-time, igniting panic and making the worst-case prophecies self-fulfilling. The threat of 2012 could be used as a pretext for drastic action, such as rushing dangerous drugs to market because there might be a plague on the way, or diverting precious trillions to an untried asteroid defense system. Worse, the growing expectations of global catastrophe in 2012 might be seized upon by malefactors who fulfill that expectation by perpetrating their own brand of mayhem.

Who would benefit if 2012 indeed turns out to be a year of chaos? Terrorists, One World oligarchs, or both, acting in concert?

"Apoca-freaks," as I have come to call them, are an occupational hazard of the 2012 business. I have encountered scores of them in three years of researching, writing, and speaking on the topic. Apoca-freaks are stimulated by the prospect of global cataclysm and flock to the 2012 concept in search of self-importance. Empowered by the belief that civilization is going to hell in just a few short years, they possess the most potent Truth in the world, and can therefore leapfrog to the top of the human heap, past all those who are more successful, more respected, more beloved. Indeed, they may succumb to a 2012 frenzy, a hysteria whipped up by the same kind of pernicious nonsense-spewers whom Albert Camus ridiculed in *Caligula*, a book about the Roman emperor who grew insane because he could not possess the Moon, and who justified his sadistic, murderous appetites by declaring, "Death exists, therefore life is absurd."

Twenty-twelve exists, therefore today is absurd?

Expect absurdism to make a comeback. It effectively relieves the flabbergasted dismay one feels as one contemplates the approach of megadoom. Several months ago, my son's preschool teacher informed me that she had upbraided him for excessive free association; when asked his favorite ice cream flavor, he looked around and said "building," causing the other four-year-olds to giggle uncontrollably and shout their own nonsense answers, including, as I recall, "toes," "running around and around," and, even less explicably, "no thanks." The essence of their teacher's objection was that preschoolers must provide sensible responses to the questions they are asked, and that any student doing otherwise threatens, however hilariously, the order and discipline of the classroom.

God willing, nonsense epidemics will prove the greatest challenge we face. One is compelled to observe, however, that nonsense is a lot funnier under the influence of drugs. Karl Marx famously observed that when a theory grips the masses, it becomes a material force. If the emerging theory today is that the world is in for a wallop, people are going to take whatever it takes to ease that pain.

Common sense says that the odds of some mind-boggling mega-cataclysm befalling the world during our lifetime seem to be less than half, but how much less? Not enough, I firmly believe, that we can ignore the possibility, not with the way economic, ecological, political, religious, and even extraterrestrial forces, such as the increasingly erratic behavior of the Sun, are converging upon us multiplicatively today.

What I fear most is the domino effect. But the good news about the domino effect is that all you have to do is shore up one of them to keep the rest from falling. Quite the opposite of a self-fulfilling prophecy, warnings about what might well happen in 2012 can and must serve to identify potential weaknesses in our defense. Consequently, one of the key concerns of this book will be to identify the dominoes that we can keep upright. Disrupting the famine-plague continuum, for example, is largely the purview of the World Health Organization. What is WHO's Code Blue scenario for a global pandemic or, worse, a polypandemic such as the seven plagues foretold in the book of Revelation? As explored further on in these pages, WHO has a comprehensive strategy to identify and bolster critical chokepoints in the famine-plague continuum, to keep the hunger domino from falling over into plague, though the organization does not have nearly the financial and legal resources to implement its lifesaving plan.

ANCIENT WISDOM, MODERN SCIENCE

Some coincidences just insist on being noticed. In the New Testament book Acts of the Apostles, Peter instructed those who lived in Jerusalem to heed what the prophet Joel had to say:

God says, "This will happen in the last days; I will pour out upon everyone a portion of my spirit; and your sons and daughters shall prophesy; your young men shall see visions, and your old men shall dream dreams. Yes, I will endue even my slaves, both men and women, with a portion of my spirit, and they shall prophesy. And I will show portents in the sky above, and signs on the earth below—blood and fire and drifting smoke. The sun shall be turned to darkness and the moon to blood, before that great, resplendent day, the day of the Lord, shall come. And then, everyone who invokes the name of the Lord shall be saved." (Acts 2:17–21)

Lord knows there are plenty of people prophesying doom these days. And as for the "portents in the sky above," it turns out that on November 13, 2012, there will be a total solar eclipse, indeed "turning the Sun to darkness," one week after the presidential election on November 6, 2012. Later in the month, on November 28, there will be an extremely rare eclipse of what is known as the Blood Moon, so named because it sometimes shows red in the sky. The Blood Moon is also known as the Hunter's Moon, reflecting the final slaughter of the year.

For 2012 to have an aftermath, as this book presupposes, apocalypse cannot, by definition, physically have occurred, at least not in the usual, final-curtain sense of the word. Rather, 2012 seems destined to fulfill the original Greek root meaning of "apocalypse," the "lifting of the veil," yielding an aftermath in which the great and terrible truths that underlie our existence are revealed.

No single discipline or approach can be counted upon to yield a reliable assessment of the future, particularly a future that threatens to be such a precipitous break from the status quo. Therefore, I have sought out the opinions of scientists, intellectuals, shamans, seers, priests, and artists, and not just in the United States, which, as political, economic, cultural, and, more often than not, moral leader of the world, is heavily invested in maintaining the global status quo. The very idea of apocalypse is an affront to those in charge. For that reason, my research into 2012 has also taken me to Asia, Europe, Africa, and Latin America.

The gold is found wherever ancient wisdom connects with modern science. The heroes of 2012 will be like Salah Kathalay, a spear fisherman of the Moken sea gypsies in Thailand, who saved his people from the terrible Indian Ocean tsunami that killed approximately 250,000 people, destroying villages and towns throughout the region. As reported on 60 Minutes, the venerable CBS newsmagazine, the Moken are perhaps the most amphibious people in the world, swimming, fishing, and boating on a daily basis, and from a very young age. On that fateful day, December 26, 2004, Kathalay methodically, one might say scientifically, observed changes in tidal patterns, in the behavior of animals and insects living near the shoreline, and quickly decided that something was up. But how did he know that it was a killer tsunami? Kathalay kept faith with the ancient Moken oral tradition that spoke of the Great Wave that eats human beings, and persisted in his warnings despite being dismissed by most of his cavalier fellow villagers as an elderly, inconsequential man living in the past. Kathalay refused to be ignored, demanding that the Moken scale the hills, which they did just in time, ahead of the killer waves even as the people of similar coastal cultures in the region, who had the same measure of practical knowledge but had lost their sense of tradition, perished in the flood. If Kathalay's warnings had proved groundless, so what? A day or two spent climbing unnecessarily up and down a hill.

As explored in my previous book, Edgar Cayce (1877–1945), perhaps the most famous psychic of the twentieth century, would go into a trance, receive visions of the future, and then write them down when he woke up. In 1934, Cayce foresaw, in this (scientifically inexplicable) manner, the global warming threat, predicting massive changes in the Arctic and Antarctic, and the subtropical warming of normally temperate areas. A world-renowned healer, Cayce perhaps somehow "felt" that the Earth had a fever decades before scientists noted the phenomenon. Even if scientists of that era suspected that global warming was occurring, they could not have expressed such an opinion professionally because, in the 1930s, there was insufficient atmospheric temperature data to support any such hypothesis. By the way, in another well-documented session, Cayce foretold a dramatic shift in the Earth's magnetic field, with cataclysmic consequences for the planet's surface.

Mitar Tarabic (1829–1899), an obscure nineteenth-century Serbian peasant, not only predicted the warfare and turmoil of the twentieth century, but pegged the aftermath as well: "You see, my godfather, when the world starts to live in peace and abundance after the second all-out war, all of that will just be an illusion, because many will forget God, and they will worship human intelligence," declared Tarabic, as recorded in The Balkan Prophecy by Zoran Vanjaka and Jura Sever. Ironic that after decades of the most gruesome carnage the world has ever seen, we should think we are so smart. Tarabic also foresaw "a kind of gadget with images," "airships with guns falling from the sky," and that men "will travel to other worlds to find lifeless deserts there and still, God forgive him, he will think that he knows better than God himself."

Like Cayce, Tarabic made no specific mention of 2012, but for the early twenty-first century he foresaw the meteoric rise to global stardom of a small holy man from the North. Though apparently a good man, this new messiah will be surrounded by hypocrites and evildoers who will wreak havoc on the planet, causing people to die in great numbers. The survivors will flee to "the mountains with three crosses," where they will at first find refuge but then encounter famine, with a twist: there will be plenty of food but it will be poisoned. Much like the sailors dying of thirst in the ocean in Coleridge's "The Rime of the Ancient Mariner," where there was "water, water every where, nor any drop to drink," only those who can resist eating and who hold out until the very end will survive and be close to God, according to Tarabic's vision.

"When wildflowers lose their fragrance, when grace leaves man, when rivers lose their health, then the greatest all-out war will come," predicted Tarabic, adding that the war would be fought with "cannonballs that do not kill, but that cast a spell over the land, the livestock and the people."

THE ODDS ARE STILL WITH US

My previous book on this subject concluded that the evidence pointing to 2012 being profoundly pivotal and quite possibly catastrophic is too eerily compelling to ignore. To those who point out that doomsayers have been predicting catastrophe forever and yet here we still are, I say, "From your lips to God's ears." I'd much rather be wrong, and for all of us to go on enjoying life much as we do now.

Odds are that the optimists are right, that 2012 will come and go much like any other year, that magazines will still be filled with photos of pretty girls, that cigarettes will still be bad for you, that Subway will still sell its Footlongs, that Christmas will still be stressful and merry. But just how high do the odds have to be in our favor to ignore the other, much more tumultuous possibility? Let's say, for example, there were a one-in-a-hundred chance that the roof of your house was going to collapse in the next five years, and, to extend the analogy to living on Earth, you were unable to move to another house. Would you prepare yourself and your family for that possibility with prayer, crash helmets, suicide pills, or Mai Tais? How about if the odds were one in ten? One in three?

For my part, I feel as though I'm one of the building inspectors who detected the potentially fatal design flaw, not just in my roof, but in everyone's. Think of *Aftermath* as a how-to guide for patching up the roof, and for what to do in case the ceiling really does start coming down around our ears. And perhaps also as a beginner's gazetteer for glimpsing through the cracks and holes at the heavens above.

"It was the best of times, it was the worst of times, it was the end of times."

Is this how Charles Dickens would have opened A Tale of Two Cities, had he somehow written it about 2012, rather than the French Revolution of 1789? That, in a nutshell, is the thesis of this book—that history will prove 2012 to be of such great importance as to start Time all over again, warranting a whole new calendar, beginning at Year One. Of course, Year One was also declared after the French Revolution. Post-1789 calendars featured ten-day weeks and were used for more than a decade until Napoleon Bonaparte restored the original Gregorian system.

Whether 2012 will reconfigure time permanently or simply shake it up for a while, one can only guess. To get a better handle on the potential significance of the date, it is helpful to examine other legendary and/or infamous points in history, such as Y2K, 1/1/01, to which 12/21/12 is often compared. Y2K was a dud, and I, for one, wouldn't mind a bit if the fabled Mayan end-date turned out the same way. Better a little egg on the face than a world that cracks like an empty shell. But 2012 has no substantive similarity to Y2K, which focused fears over the consequences of a computer glitch. While the fears proved way overblown—remember the psychotherapists who specialized in calming pre-Y2K anxiety?—all the hoopla did, in fact, prompt major corporations to "Y2K-proof" their operations. No doubt the repair job saved us from some sticky, though probably not catastrophic, snafus.

What's in a date, anyway? If the fifteen-hundred-year-old Mayan prophecy about 2012 were off by a decade or two, would that invalidate their claim? It would certainly call into question their basic reasoning, that the completion of a great cycle on the winter solstice of 12/21/12 moves us into a new age, which transition somehow causally triggers and/or is ineluctably accompanied by profound changes in living circumstances around the globe. But better to be lucky than smart, as they say. Poignant warnings of doom are worth the anxiety they create if they help ward off the worst from ever happening. Perhaps 2012 will turn out like 1984, another fabled doom date that, thankfully, never lived up to its rats-eating-your-face advanced billing as the year when the human spirit was crushed once and for all. Were George Orwell's fears simply misplaced, or, as I believe, did his chilling depiction of Big Brother help in the struggle to deny totalitarianism its triumph?

Will I and all the others waving red flags about 2012 feel like fools if the year comes and goes unapocalyptically? Depends on just how far off our predictions turn out to have been. Had a comparable group of commentators warned in advance that September 11, 2001, would be the end of the world, that prophecy would of course have been proven wrong, but the apocalyptic aura of the events of that day would, in retrospect, acquit the prophets as having been onto something. Same would have gone for doom prophecies specifying the date June 28, 1914, the day on which Archduke Franz Ferdinand and his wife, Sophia, Duchess of Hohenberg, were assassinated as they entered the City Hall in Sarajevo. Historians generally consider this event to have ignited World War I—not the end of the world, exactly, but the closest we had ever come.

Of course, 12/21/12 differs from 9/11 and the detonation of World War I in that those dates were largely unanticipated, except in certain dark corners of the world. There was precious little time to prepare. How foolish would it be for us to fritter away the valuable lead-time that the Maya have afforded us. Forewarned, as the saying goes, is forearmed, but only if the warning is taken seriously. At bottom, the Mayan warning about 2012 is not to miss the boat to enlightenment, not to miss the chance to sail above all the greed, violence, and insanity that will crest like a bloody tsunami breaking over 2012. In that year, humanity will begin to move into a new age, just as in 1492, humanity, or at least its largest and most dominant group, sailed into a New World.

James Reston, Jr., writes in Dogs of God: Columbus, the Inquisition, and the Defeat of the Moors:

> "Fourteen hundred ninety-two is a year that can aptly be called apocalyptic both in the original meaning of the word, as revelation or disclosure, and in its more modern usage of colossal calamity. That so many important forces of history converged at one time inevitably begs the question of whether the hand of God was at work in the confluence. To the Christians, the Arabs, and the Jews of the late fifteenth century alike, there was no doubt. Such great and terrible things do not happen simultaneously at random. Providence had to be involved, and the major players were merely God's instruments, either for glory or for disaster."

Reston powerfully demonstrates that 1492 was so important and complex that it still stirs debate half a millennium later. Undeniably, it heralded the greatest expansion of civilization in history, and just as undeniably, it initiated an epoch of brutal slaughter, not just of many millions of people, but of a harmonious indigenous relationship with nature whose lost lessons we are desperately in need of today. No surprise that the Maya and other native Americans tend to view that year much less enthusiastically than most other folks, less as the discovery of a New World than as the end of an old and dear one.

AFTERMATH

One thinks of the Spanish Inquisition, 1478 until 1530 and beyond, as something that happened in Spain to Jews, Muslims, and, to a lesser extent, Protestants, but in point of fact, many of the conquistadors who invaded the Mayan kingdom in 1519 were bloodlust veterans of the persecution and genocide.

"As if it weren't enough, Catholic priests, most of whom were inquisitors, accompanied the conquistadors. These holy men saw the devil at every turn and strove to destroy all traces of the great Mayan culture. They murdered our Elders and H-Menob', or Mayan priests, leaving people devastated and deprived of their leaders. The Elders knew this period of darkness was coming because the prophecies had been very clear. Therefore, many of them had already moved deep into the jungle or hidden in the highlands, where they maintained their lineages, rulers and cosmovision," writes Barrios. (In deference to the Mayan predilection for numerology, it should be noted that 2012 comes 520 years after 1492, 520 being precisely one-tenth of their 5,200-year-long cycle, one-fiftieth of the aforementioned 26,000-year cycle, and also ten times the number of weeks in the year, 52, which they regard as a number of significant power and felicity.)

Maybe 2012 will ultimately prove the karmic flipside of 1492, that is, an exhilarating reawakening for indigenous cultures and a god-awful setback for the rest of us. The question of whether or not this sort of poetic justice operates in history is beyond the scope of this book. A much simpler scenario yielding the flipside outcome whereby indigenous folks gain and the sophisticates lose has less to do with justice than electricity. The more a society depends on electricity, and on the technological infrastructure based on that form of power, the bigger the bashing it will receive in 2012.

HERE COMES THE SUN

althought the date would stump most trivia buffs, September 2, 1859, is when the greatest magnetic storm ever recorded hit the Earth. It is also the date likeliest to be replayed in 2012, with one important difference: this time, the devastation will be colossal.

The Carrington event, named after Richard Carrington, the amateur British astronomer who took the lead in observing and explaining it, was actually a one-two punch that uppercut the Earth over the course of a week. The first of the two massive solar explosions began forming sometime in mid-August 1859, when an unusually large sunspot appeared on the northwest portion of the Sun's face. On August 27, it erupted like a zit, shooting out a Moon-sized glob of plasma, or supercharged gas. Such blasts are known as coronal mass ejections (CMEs).

CMEs are usually shaped like croissants, according to a discovery made in 2009 by STEREO, a pair of NASA probes that flank the Sun and photograph these explosions from opposite sides. According to Angelos Vourlidas of the Naval Research Laboratory, a computer model designer for the STEREO mission, CMEs are formed in a manner akin to that of twisting the ends of a rope around and around, tighter and tighter, until the middle bulges out. Instead of rope, Slinky-like lines of magnetic force twist out of the sunspots. Eventually, after enough twisting, the crescent-shaped coil of plasma snaps free and spins away from the Sun at a million miles per hour or more, which is just what happened in the Carrington event.

The first cosmic croissant of the Carrington event hit Earth the next day, August 28, 1859, causing some of the most beautiful auroras ever seen. The northern lights don't normally extend down to Havana, Cuba, but this time they did, making the sky there appear as though it were stained with blood and on fire.

On September 1, 1859, the Sun erupted again, even more furiously. According to scientists' reconstructions, the second Carrington CME was dozens of times more powerful than average, weighing in at about 10 billion tons and 10 trillion trillion watts (trillions of times more than the sum total of all electrical, mechanical, combustible, muscular, animal, and plant energy than has been produced or consumed in the history of the planet). Traveling at about 5 million miles per hour, it was also one of the fastest ever recorded. Think of a tennis ball machine suddenly rifling out a (molten, radioactive) basketball.

When CMEs launch, they create a shockwave that slaps the solar wind, a sphere of charged particles, mostly protons. This impact causes what is known as an SEP (solar energetic particle) event, which accelerates everything in its path exponentially; most of these supercharged particles take an hour or less to reach the Earth's atmosphere, where they fuse nitrogen and oxygen atoms to create nitrates, which eventually settle as dust onto the poles. Although the Carrington SEP is generally considered the largest on record, back then no one noticed it because there were no instruments sensitive enough to detect it. (Evidence of the 1859 SEP impact has since been found in anomalous nitrate-laden ice core samples that date back to that time.) Today, there are satellite-borne instruments sensitive enough to detect SEPs, most of which would probably have been fried by the Carrington event's ferocity. Indeed, far lesser SEPs are blamed for having disabled a number of spacecraft, including Japan's Nozomi satellite, dooming that nation's mission to Mars. SEPs also threaten astronauts; a Carrington-scale event would imperil those aboard the International Space Station.

At 4:50 GMT on September 2, 1859, the second and by far the more powerful Carrington CME barreled into the Earth, fifteen to twenty hours behind the SEP shockwave it had detonated. The CME made quite a splash in the headlines, sizzling telegraph wires, causing fires, and filling the sky with an auroral glow that made midnight as bright as noon.

"The electricity that attended this beautiful phenomenon took possession of the magnetic wires throughout the country, and there were numerous side displays in the telegraph offices where fantastical and unreadable messages came through the instruments, and where the atmospheric fireworks assumed shape and substance in brilliant sparks," reported the *Philadelphia Evening Bulletin*. The electrical blasts were so powerful that some telegraph operators disconnected the batteries to their equipment and were still able to send and receive messages just operating on the power that was heavenly supplied.

Were we hit today by a geomagnetic storm of equivalent strength to the Carrington event, our civilization could well be plunged into chaos. This is not an exaggeration. Rather, it is the consensus of those who presented at the National Academy of Sciences' report *Severe Space Weather Events: Understanding Societal and Economic Impacts*, published in December 2008. The report's executive summary says:

> Because of the interconnectedness of critical infrastructures in modern society, the impacts of severe space weather events can go beyond disruption of existing technical systems and lead to short-term as

well as to long-term collateral socioeconomic disruptions. Electric power is modern society's cornerstone technology, the technology on which virtually all other infrastructures and services depend... Collateral effects of a longer-term outage [such as would almost certainly result from a Carrington-scale space weather event] would likely include, for example, disruption of the transportation, communication, banking, and finance systems, and government services; the breakdown of the distribution of potable water owing to pump failure and the loss of perishable foods and medications because of lack of refrigeration. The resulting loss of services for a significant period of time in even one region of the country could affect the entire nation and have international impact as well.

Contributors from NASA, NOAA (National Oceanographic and Atmospheric Administration), the Smithsonian Institution, the United States Air Force, a number of major universities, and advanced technology corporations gave evidence that a contemporary Carrington-scale event would lead to deep and widespread social disruption. Basic to this contention are the enormous changes to the United States' infrastructure over the past century and a half. Modern society is utterly dependent on electricity. The electrical system is the master system upon which all others depend. And it is vulnerable to historically large space weather events.

"Emergency services would be strained, and command and control might be lost," concludes the committee of National Academy of Sciences researchers, chaired by Daniel Baker, director of LASP, the Laboratory for Atmospheric and Space Physics, at the University of Colorado, Boulder.

Baker's concern about the consequences of space weather is quite a turnabout for LASP researchers. Readers of my previous book might recall the part where I attended a solar physics conference in Colorado sponsored by LASP, only to find that the scientists assembled there were utterly indifferent to a space weather freak-out occurring even as they met. The week of September 7–13, 2005, right after Hurricane Katrina and just before Rita and Wilma, goes down as one of the stormiest periods ever recorded on the Sun, but at the LASP conference, which began on September 13, no one even mentioned this astonishing situation, not even during the coffee breaks.

What no one at LASP or any other space laboratory has ever disagreed with, however, is that the fiercest solar storms usually occur at the climax of the eleven-year solar cycle, which, by general scientific consensus, is next due in late 2012 or early 2013.

SPACE WEATHER BLUES

With so much hanging in the balance, one might think there would be legions of space weather experts scanning the sky for signs of impending catastrophe, that the best and the brightest would be lining up for a chance, quite literally, to save the world. But much of the talk at the May 2008 workshop that gave rise to the National Academy of Sciences' report was about how hard it is to get people interested in space weather. Students don't sign up for the classes, and when, in rare instances, such coursework is required, their eyes glaze over, according to Paul Kintner, professor of electrical and computer engineering at Cornell University.

The air force, responsible for all U.S. assets in space, has tried to overcome this indifference by offering extended space weather education at air force expense, but the number of expert space weather forecasters has nonetheless declined steadily.

"The DOD is striving to increase the sampling of the space weather environment for the coming solar maximum [in 2011–2012] and beyond," says Major Herbert Keyser, United States Air Force Weather Agency.

However, he also notes that expertise in space weather is a national resource that is quickly disappearing.

European efforts aren't going any better, their space weather activities being described as "complicated" and "highly fragmented." Russia has a creditable program, as do China, India, and Japan, though most of this effort seems oriented toward their respective space programs rather than protecting us here on the ground.

Why so ho-hum? For one thing, there's almost no budget. The world's principal supplier of space weather information, the Space Weather Prediction Center (SWPC), operated by NOAA, has what was referred to as an "unstable budget of $6 to $7 million per year." True, the SWPC shares resources with NASA and also the Air Force Weather Agency (AFWA), but still, given the stakes involved, it's a piddling amount. The deeper reason, one suspects, is that we have not really gotten whacked yet, not hard like the way an 1859 or 1921 storm would, in the Internet Age, be the glitch to end all glitches, and perhaps even to end any memory of what a computer glitch ever was.

"It was lamented that, in the eyes of the public and policy communities, severe space weather lacks salience as a problem; it is very difficult to inspire non-specialists to prepare for a potential crisis that has never happened before, and may not happen for decades to come. Attention is inevitably drawn toward higher-frequency risks and immediate problems," says the National Academy's report.

Damn those credit default swaps! Non sequitur? Not really.

Founded by President Abraham Lincoln during the height of the Civil War, the National Academy of Sciences serves as America's, and indeed often the world's, highest court of scientific opinion. There seems little doubt that the compelling evidence presented in the National Academy of Sciences' report would, at almost any other time, have had a real chance of heightening awareness of and funding for protective space weather research, and perhaps also mitigation programs to harden and defend the power grid, satellite system, and other vital assets vulnerable to solar predations. But between the time the workshop was held in May 2008 and the proceedings were published later that December, Wall Street fell through the floor, sucking a trillion or two tax dollars down with it. The economic crisis was so sudden and severe that the public and their legislators seemed to get "crisis overload," relegating all other matters, regardless of how urgent, to the back burner. True, the federal government's stimulus program could have shoveled some funds to space weather protection services, but it appears that the emergency monies directed toward NOAA and NASA have gone to other programs instead.

The greatest peril posed by the credit-crunch stock market crash that began in September 2008 and the subsequent Herculean investments made to defibrillate the global economy is that it fixates us on what's most urgent instead of what's most important. During crisis times, it seems that the future had damn well better take care of itself.

DEEP SOLAR MINIMUM

On its face, the decision to forego funding space weather expansion in 2009 was a calculated risk. One the one hand, the Sun has been unusually active over the past few decades, with numerous aberrant events.

"Since the Space Age began in the 1950s, solar activity has been generally high. Five of the most intense solar cycles on record have occurred in the last fifty years," said solar physicist David Hathaway, a veteran Sun watcher from NASA's Marshall Space Flight Center in Huntsville, Alabama.

AFTERMATH

However, the period from 2008 through much of 2009 saw a sharp fall-off in solar activity. The simplest and most common way to gauge solar activity is by sunspots, which astronomers in China have counted for the past two thousand years; actual drawings of sunspots made by Galileo are kept in the Vatican archives. Sunspots are planet-sized magnetic storms on the surface of the Sun. They are the source of CMEs, such as the one that caused the Carrington event, as well as the source of most solar flares and intermittent blasts of ultraviolet radiation. By this measure, 2008 was virtually dormant; almost three-quarters of its days had no sunspots. One had to go back to 1913 to find a less active Sun. During the first quarter of 2009, almost 90 percent of the days were sunspotless, a historic low.

"We're just not used to this kind of deep calm. This is the quietest Sun we have seen in almost a century," adds Hathaway.

Was the Sun finally cooling its jets? Radiowave emissions had dropped to their lowest levels in half a century, perhaps indicating a weakening of Old Sol's magnetic field. This would jibe with the fact that the solar minimum of 2008–2009 also saw a record low in solar wind pressure, the lowest since measurements began to be made in the 1960s. Solar wind is composed of elementary particles such as protons and electrons; the fewer the particles, the lower the pressure. Measurements from Ulysses, a NASA solar research satellite, further indicated that the Sun's brightness, known as solar irradiance, had hit a twelve-year low; visible wavelengths had declined by 0.2 percent since the solar minimum of 1996, with ultraviolet wavelengths plummeting 6 percent.

With a calming Sun, who could blame embattled policy makers trying to fight off an economic depression for gambling that the next climax will be less of a bang than a whimper? The solar minimum of 2008–2009 was so low that on September 22, 2008, NASA held a special teleconference on the subject. A panel of experts presented and assessed the data concerning these anomalous solar conditions. Sadly, there were no climatologists on the NASA panel, and none of the members would comment on the implications of this change in solar wind for the Earth's well-being. But clearly, a reduction in the power of the coming 2012 climax would be welcome relief. So the question on everyone's mind was, is this a trend, or a blip?

"Usually after a long, low solar minimum, the following maximum is steep and intense," said Karine Issaultier, a solar physicist with l'Observatoire de Paris, Meudon, who was one of the panelists in the NASA teleconference. When I called Issaultier in France several days later, she expounded on her misgivings. "This solar minimum does not fit the classical model," said Issaultier. She explained that the 2008–2009 solar minimum, though lacking in sunspots, is otherwise surprisingly active, with immense streamers and prominences decorating much of the face of the sun. "I don't know why they are so large," she said.

The day after we spoke, the largest solar prominence in years, ten times the size of Earth, erupted magnificently.

Currently, solar physicists are split into two camps regarding the solar climax of 2012. As examined in my previous book, some studies, including the landmark research done by Mausumi Dikpati and her colleagues at the National Center for Atmospheric Research's High Altitude Observatory, anticipate that this climax will be 30 to 50 percent greater than ever recorded. Such an outsize climax would unleash monstrous storms and explosions. Other researchers, including Hathaway and Dean Pesnell, a solar physicist at the Goddard Space Flight Center, anticipate a below-normal peak coming in 2012–2013. Complicating matters even more is the fact that strong flares can occur during weak solar cycles. In fact, according to Tony Phillips, editor of SpaceWeather.com, a NASA website, the 1859 Carrington event, considered the granddaddy of all recorded solar blasts, "itself occurred during a relatively weak cycle similar to the one expected to peak in 2012–2013."

GAPING HOLES

What if you suddenly discovered that the door to your home had been kicked in and you could neither close nor repair it? A storm is kicking up and blowing all sorts of nasty things in, and unless you're mistaken, it seems as though your house is actually sucking stuff in as well. I couldn't get these images out of my head, so I called my friend Dr. Mary Long, a psychologist in Bellport, New York, on the south shore of Long Island, and ran it by her.

"Sounds like you must be going through a very tumultuous time and that maybe there's a little paranoia involved here," said Dr. Long.

"What about if I told you that, instead of just my house, this is happening to the entire planet?"

"Then I'd say you're having delusions of grandeur, imposing your personal problems on the world at large," replied Dr. Long.

So it's all in my head . . .

But picture this: instead of the front door, the door to the Earth's protective magnetic shield has been kicked in. That, in essence, was the discovery made in 2008 when THEMIS, a fleet of five NASA unmanned spacecraft, accidentally flew through an immense pole-to-equator breach in that field. This breach extends out into space to create a hole in our planet's defenses four times as wide as the Earth itself, and ten times larger than any such hole ever detected or even theorized. Consensus among the space physicists analyzing the data was that this megachink in Earth's armor should make for the strongest geomagnetic storms ever, storms that will blast us with doses of solar radiation that will fry our eyes, our skin, and our infrastructure—everything from our electrical power grids to the global satellite network that runs our telecommunications, commerce, military security, and law enforcement.

"This kind of influx [of potentially dangerous solar wind particles blowing through this gigantic new hole in our protective magnetic shield] is an order of magnitude greater than we ever thought possible," reports Wenhu Li, a space physicist at the University of New Hampshire who was part of the team that analyzed the NASA data.

"The more particles, the more severe the storm," says Li's colleague Jimmy (Joachim) Raeder. "If the solar field has been aligned with the Earth's for a while, we now know Earth's field is heavily loaded with solar particles and primed for a strong storm . . . In fact, we expect stronger storms in the upcoming cycle. The sun's magnetic field changes direction every cycle, and due to its new orientation in the upcoming cycle, we expect the clouds of particles ejected from the sun will have a field which is at first aligned with Earth, then becomes opposite as the cloud passes by." Raeder adds that the anatomy of this particular breach makes it susceptible to CMEs that occur during even-numbered solar cycles, such as the current cycle, Cycle 24, expected by resounding scientific consensus to peak late in 2012.

"It's the perfect sequence for a really big event," Raeder concludes.

To understand why the Earth will be so susceptible to CMEs emitted during the 2012 solar climax, try this little thought experiment. Take two magnets, one very large, representing the Sun, and one small, representing the Earth. Place them so that their poles are aligned, north to north or south to south. In this position they are repelling each other and therefore are not touching. This position is analogous to how magnetic fields of the Sun and the Earth are currently aligned, north to north.

The force field between the two aligned poles is in fact made up of zillions of charged particles. Until the THEMIS discovery, scientists believed that when the Sun and the Earth were in a north-to-north or south-to-south position, the danger to Earth from solar blasts was relatively small, because the blasts, having the same magnetic polarity as the Sun from which they came, would therefore be repelled. Conversely, the danger to Earth was believed to be much greater when the poles are oppositely aligned, conditions in which supercharged particle blasts from the Sun would get magnetically sucked right into the Earth's outer atmosphere, occasionally resulting in blackouts, satellite failure, and other such disturbances, including those caused by the Carrington event of 1859. Note that as recently as the publication of the National Academy of Sciences' *Severe Space Weather Events* study in December 2008, this was the prevailing orthodoxy.

Now the astrophysicists think differently. Imagine that the two magnets have been held north to north for years, just as the Sun and the Earth have been aligned. For all this time, the Sun's magnet has been emitting a stream of charged particles toward the Earth's. The force field between them has grown stronger and stronger. The THEMIS astrophysicists now believe that sometimes the pressure of this powerful force field becomes so great that the Sun literally rips away some of the Earth's magnet's protective field, allowing in a massive flood of those stored-up charged particles. If the door won't open, then kick it in.

Researchers had theorized the existence of such a "closed door" entry mechanism, but had no idea how important it was, according to Marit Oieroset, a geophysicist at the University of California, Berkeley, who is one of the lead researchers on the THEMIS project.

"It's as if people knew there was a crack in the levee, but they did not know how much flooding it caused . . . Twenty times more solar particles cross the Earth's leaky magnetic shield when the sun's magnetic field is aligned with the Earth compared to when the sun's magnetic fields are oppositely directed," says Oieroset. Brute-force entry is now implicated in the largest CME episodes, including, quite possibly, the Carrington event of 1859.

Now for the absolutely worst-case scenario, which is what we may well be looking at in just a few short years. Take the Sun magnet (which has been held in place for years) and quickly flip it around to the other pole. This is what the Sun is widely expected to do in 2012, to reverse polarity, as it does at the climax of every second solar cycle, every twenty-two years. The magnets would swiftly snap together. The force field that had built up between them would zap the smaller Earth magnet.

Why, you might ask, wouldn't it flow back toward the larger, more powerfully attractive Sun magnet? Actually, it would, in our simple experiment, but not so in the real-world case of the Sun and the Earth. The reason for the difference is that the power of the magnetic force between two sources is inversely proportional to the *square* of the distance between them. The same basic inverse square principle applies to gravity; recall that the gravitational balance point between Sun and Earth, called the Lagrangian point, is about 1 million miles out from our planet. The electromagnetic force field built up around the Earth would obviously be very close to our planet, much less than one million miles away. Thus, the pent-up radiation trapped north to north just outside our planet will bolt toward us, not toward the Sun, if the Sun's polarity flips to create a north-to-south alignment, as is roundly predicted to happen in 2012. Put another way, the door will fly open in 2012 and our house will suck in the radiation cloud hovering outside.

Personally, I feel that this unfathomably immense hole in our planet's protective magnetic field is somehow connected to the deep solar minimum of 2008–2009 and the concomitant reduction in the Sun's magnetic field. True, our magnetic field emanates from the Earth's molten core, not the Sun. But the two magnetic fields do meet in outer space, where they interact energetically. Perhaps the dramatic reduction in the Sun's energy output during the recent solar calm somehow elicited a reciprocal reduction in the Earth's magnetic output. Such a theory would imply that when solar output resurges, so will the strength of the Earth's protective magnetic shield. I know it's just a lay hypothesis, but if it's wrong, the net-net is not at all good.

"The [THEMIS] discovery overturns a long-standing belief about how and when most of the solar particles penetrate Earth's magnetic field, and could be used to predict when solar storms will be more severe. Based on these results, we expect more severe storms during the upcoming solar cycle," says Vassilis Angelopoulos of the University of California, Los Angeles, principal investigator for NASA's THEMIS mission.

progress makes us stronger, right? In most ways, it does. But the paradox of progress is that it fosters dependency. If, for example, all of the automobiles in the world stopped running, we would be in a much worse situation than if there had never been any automobiles in the first place, because we have designed our lifestyles around the automobile's capabilities. Living three miles east of the supermarket, seven miles south of the hospital, and twelve miles west of work, ordinarily a no-brainer, can become a huge problem without a car.

Nowhere is the paradox of progress more extreme than in the case of the power grid. With it, we are immeasurably more able than our pre-electric ancestors: we can travel, communicate, enjoy, attack, and defend in ways they never could have imagined. But without the magic juice, we are defenseless.

So how did we get hooked?

When the Carrington event hit in 1859, George Westinghouse was eleven and Thomas Edison was twelve, good ages, one might suppose, for that magnificent burst to have sparked the budding young inventors' imaginations. Edison and Westinghouse went on to design and build power grid systems that quite literally electrified the United States and soon much of the rest of the world.

AFTERMATH

In what is known as the "War of the Currents," Westinghouse, backed by the Serbian genius Nikola Tesla, battled Edison to see which system would prevail. Westinghouse's system was based on generators that used whirling armatures to produce alternating current, an electrical flow that reverses direction twice at a rate of sixty times per second. Edison's system relied on batteries that discharge direct current (DC), which does not reverse itself and thus goes only in one direction.

Alternating current (AC) can be more easily transmitted over long distances, and can be readily stepped down at the point of reception, allowing it to be divided among various appliances that would use it. Just one line of current can power multiple applications. Direct current cannot be transmitted over such long distances, requiring that power plants be built much closer to the end-users. This required greater capital investments and also greater intrusion into residential areas. DC transmissions likewise cannot be stepped down nearly as conveniently as alternating current, meaning that a separate direct current power line would have to be run into the house for each appliance. (Imagine how many separate power lines it would take to supply today's typical household, all its kitchen appliances, lighting fixtures, computers, and televisions.)

Edison's system, though clearly inferior, almost prevailed because of his personal fame and the flamboyant, often dirty campaign he waged on its behalf. Direct current's downfall came, quite literally, with the downfall of the great tangle of power lines required to supply it. The Great Blizzard of 1888 collapsed many of these lines under the weight of ice and snow, causing numerous electrocutions. That tragedy is cited as the single most important event in turning public opinion against Edison and in favor of the superior Westinghouse/Tesla AC system.

Had the dates been switched, had the Great Blizzard come in 1859 and the Carrington event in 1888, the electrification of society would have proceeded much differently. By 1888, Edison had at least 121 DC electrical systems operating in the United States, most of them in the Greater New York area. Westinghouse had more than thirty AC electrical systems, centered in New England. The (1888) Carrington space weather blast would have come, as all space weather events do, in the form of immense amounts of DC electricity, current flowing directly from the Sun to the Earth. Edison's system could have handled the overload, only a couple of extra amps of direct current, without much problem, but Westinghouse's system, the one still in use around the world today, would have shorted out completely.

In all likelihood, the Westinghouse system, superior in every other respect, would have still won out in the end, but the spectacular blowout would have alerted engineers from that point on to look for ways to protect our electrical power grids from another such catastrophe. We would have had 130 more years of ingenuity applied to protecting our electrical grid from a space weather blowout than we do now.

ACED OUT

On March 13, 1989, two solar blasts, each about a tenth the size of the Carrington blasts, knocked out the Hydro-Québec electrical utility, causing it to go from fully operational to complete shutdown in ninety-two seconds. On a computer simulation of the event, the blast looks like a giant red, toothy mouth taking bites out of the top of the Northern Hemisphere. Millions of customers in Quebec and Scandinavia lost power, but within nine hours it was restored. No big deal in the grand scheme of things. True, a number of nuclear-, oil-, and coal-powered plants as far away as Los Angeles subsequently reported transmission anomalies, but nothing blew up.

One might even say that March 13, 1989, served as a wake-up call. At the time, we were not able to measure the solar wind, meaning that we had precious little ability to predict when the next CME was coming, and no time to prepare for its impact. In a follow-up study, Oak Ridge National Laboratories determined that a storm only slightly larger than that which occurred in 1989 could result in a $36 billion economic loss, not including collateral losses to critical services such as transportation, fire protection, and public security. So NASA got to work and in 1997 successfully launched Advanced Composite Explorer (ACE). ACE sits in the Lagrangian point, also known as L1, a spot about a million miles (1.5 million kilometers) away where the Earth's gravitational field balances with that of the Sun. There, nestled into what is known as a gravity well, ACE flies endlessly around in a tight little circle, called a "halo orbit," which serves to minimize the Sun's radio noise so that its own transmissions won't be drowned out.

"ACE has a prime view of the solar wind, interplanetary magnetic field and higher-energy particles accelerated by the Sun, as well as particles accelerated in the heliosphere and the galactic regions beyond. ACE also provides near real-time 24/7 continuous coverage of solar wind parameters and solar energetic particle intensities (space weather)," write Eric R. Christian and Andrew J. Davis in *Space Science Reviews*, 86, 1, 1998.

Electricity on the power grid must be consumed the instant it is produced, meaning there's no way of storing some for a rainy, or in this case, a too violently sunny, day. Because electricity cannot be stored in anywhere near the amounts typically consumed on the grid, thousands of grid operators across North America are at work day and night to match electricity supply to demand, spot and fix trouble spots, and generally oversee the orderly transfer of power. They are kind of like air traffic controllers, except that instead of airplanes, they oversee kilovolt streams of electrical charge. According to power grid operators contributing to the 2008

National Academy of Sciences study, ACE is their most valuable space asset for predicting space weather.

"The most important device that I know of out there to give us a heads-up is ACE. ACE gives our operators about a forty-five-minute advance notice," says James McGovern, of Reliability Coordination Services, Inc. When a CME is incoming, ACE gives power grid operators just enough lead time to quick-start units, shut down vulnerable transformers, and reroute power loads. Charles Holmes of NASA Headquarters contends that perhaps the most critical weakness in the present space weather prediction system is the reliance on the aging ACE satellite as virtually the nation's only upstream solar wind monitor. ACE was designed to function for five years, but as I'm writing this it has gone for eleven years and counting. Sadly, its transmission heads are losing gain and it is running out of the fuel necessary to keep it flying in circles.

(To a certain extent, the discovery made by NASA's STEREO satellites that CMEs are crescent-shaped could offset the loss of ACE's predictive capabilities. Any crescent-shaped emanation from the Sun's surface will now be watched very closely and matched to computer-model CME archetypes to chart its development and estimate the time when it will twist away from the surface. According to NASA computer modeler Vourlidas, this will enable us to predict the time of CME impact on Earth to within three hours of uncertainty, down from twelve hours, as was formerly the case. However excellent the STEREO modeling, it still is no substitute for the forty-five-minute advance notification that ACE provides.)

At present, there are no approved plans or funds allocated to replace our ACE in the hole in the sky. Jurisdiction over the NASA satellite has been transferred to NOAA, which historically has shown much more interest in the oceanic, rather than the atmospheric, part of its mission. Should ACE burn out, as it could any day now, we would lose most of our ability to predict space weather, and along with it, that vital head start.

SPACE WEATHER KATRINA

What would you do upon learning that forty-five minutes from now the electricity was going to go out, across the nation, for months? Fly over to the store and stock up on canned goods? Gather the family in front of the TV to watch the final broadcast of CNN? Since pumps powered by electricity supply most of our water, it might be a good idea to fill up the tubs, and, oh yeah, have everyone use the restroom, since the toilet will no longer flush. But you really don't need to worry about such things, because the way things look, sooner or later—with the next attack likely to come in 2012—we're going to lose our electricity without any notice at all.

CMEs are going to knock out the power grid and bring civilization to its knees, according to numerous studies conducted by the Metatech Corporation, a Santa Barbara, California, research firm specializing in the effects of EMI (electromagnetic interference). Metatech's research, which figured prominently in the National Academy of Sciences report, used as its baseline the Great Magnetic Storm of May 1921, a CME assault that was somewhat smaller than the 1859 Carrington event and several times as large as the 1989 Hydro-Québec event. John Kappenman, senior consultant for Metatech, presents voluminous, tightly reasoned evidence that an event on the scale of the 1921 magnetic storm would today result in large-scale blackouts affecting more than 130 million people in North America alone; the Northeast, Midwest, and Pacific Northwest would be hit particularly hard because of their northerly latitudes. Multiply that misery index to encompass the devastation that would undoubtedly also befall the rest of the Northern Hemisphere, particularly Scandinavia, Western Europe, and Russia, and then multiply that total by the three or four years it would probably take for us to return to a functioning society, assuming, of course, that such a massive catastrophe had not touched off social unrest sufficient to undermine the governments and institutions in charge of the recovery. And also assuming that we weren't crippled by a follow-up CME blast in the interim.

"The experience from contemporary space weather events is revealing and potentially paints an ominous outcome for historically large storms that are yet to occur on today's infrastructure. Given the potentially enormous implications of power threats due to space weather, it is important to develop effective means to prevent a catastrophic failure. Trends have been in place for several decades that have acted to unknowingly escalate the risks from space weather to this critical infrastructure," writes Kappenman in *The Vulnerability of the U.S. Electric Power Grid to Severe Space Weather Events, and Future Outlook.*

"Historically large storms have a potential to cause power grid blackouts and transformer damage of unprecedented proportions, long-term blackouts and lengthy restoration times, and chronic shortages for multiple years are possible . . . An event that could incapacitate the network for a long time could be one of the largest natural disasters we could face," continues Kappenman, who estimates that recovering from a future severe magnetic storm would cost $1 to $2 trillion—ten to twenty times the cost of Katrina—in the first year alone. Depending on damage, full recovery from such an assault could take four to ten years, again assuming that social order had not degenerated too chaotically as a result of the traumatic infrastructure collapse.

So how does a solar blast keep your toilet from flushing? By disrupting the power grid system at its weakest point: the transformer. Transformers receive power from high-voltage transmission lines, which in turn receive their power from substations directly connected to the power plant, be it coal, oil, gas, hydroelectric, or nuclear. High-voltage transmission lines, the ones held up by those big Y-shaped metal trellis structures that can be seen stretching along the highway, carry the current as far as 300 miles. The farther the distance, the higher the voltage required, just as more water pressure would be required to produce a steady, reliable stream of water out of a long hose than out of a short one. (Volts are essentially units of pressure, while amps are units of volume. The simplest analogy is to water: volts would measure how hard the water rushes out of the hose, amps would measure how much water is flowing.) The power from the transmission lines is fed into the transformers, whose job is to then step it down from the level of hundreds of thousands of volts to tens of thousands of volts, then feed the power to be split into several directions, via a device known as a "bus," which sends the current through power lines one sees everywhere, held up by utility poles, on into homes and businesses.

Transformers operate at levels as high as 700 kilovolts (700,000 volts) in the United States and up to 1,000 kilovolts in China. Transformers in Europe typically use lower voltages, in the 400-kilovolt range. At one point, the Swedish electrical utility was considering upgrading to 800 kilovolts, but protests from groups concerned about the human health impacts of the new ultrahigh-voltage lines put the kibosh on that. Right for the wrong reason, one might observe. The higher the voltage processed by a transformer, the narrower the tolerance for error and the more vulnerable it is, therefore, to the extra electrical jolt that would come from the GICs (geomagnetically induced currents) caused by solar blasts.

A repeat of the 1921 magnetic storm would see the copper windings and leads of the 350 or so highest voltage transformers in the United States melt and burn out. Transformers weigh several tons apiece and usually cannot be repaired in the field. In fact, most transformers damaged by space weather incidents have key components fused solid, meaning they cannot be repaired at all and need to replaced with new units. Currently, the worldwide waiting list for transformers is about three years, and about half of those made fail either in test or while in service, according to Metatech's research.

But wait a minute. I understand why we are so much more vulnerable to space weather blasts than we were in the mostly preelectric age of 1859, but why so much more than the largely electrified society of 1921? The effects of the 1921 storm on society at that time were minimal, little more than the telegraph disruptions, flash fires, and eye-boggling auroras that occurred during the Carrington event. Their transformers didn't fuse, so why should ours?

I contacted Kappenman regarding this question and we arranged a meeting several days later in Los Angeles, on April 7, 2009. Kappenman, a serious and unassuming electrical engineer from Duluth, Minnesota, explained that back in 1921, there was virtually no power grid. Every city had its own coal-fired power plants, and although the systems were linked together from city to city, it was mostly for backup purposes. There was very little swapping of power from one city to the next. Rates were set by the Public Service Commission and each state had its own power utility that made sure that the system kept running.

Remember the "Electric Company" property in the board game Monopoly? It was a steady little income producer, without much risk or reward. That's the way the electrical utility industry used to see itself, safe and dependable, no hotels going boom or bust on Boardwalk or Park Place. With seemingly unlimited supplies of inexpensive coal and oil, and no competition to pressure them to upgrade or innovate, the electrical utilities settled into a comfortable, if not terribly efficient, status quo that lasted for half a century. The Arab oil embargo of 1973 quadrupled oil prices and sent everyone looking for ways to improve energy efficiency. Despite their admirable track record of reliability, the utilities were soon forced to open their transmission system to other qualified producers of electricity, much the way that Ma Bell was forced to share her lines with other telephone companies. In 1992, the definition of "qualified producers" was expanded to include just about anyone who could generate current, and a whole new entrepreneurial, and opportunistic, crop of brokers, middlemen, independent reps, and dealers invaded the marketplace and began swapping large power loads across great distances in an unprecedented effort to match supply with demand. Prices went down, efficiency increased, and so, by an order of magnitude, did the stress on the grid.

Today, the megawatt power loads that zip instantaneously around the North American power grid account for fully 40 percent of the total energy consumption in the United States and Canada. Demand for electricity is projected to grow 20 percent or more per decade, meaning hundreds of new power plants coming on line, plus many thousands more solar, wind, and other alternative energy generators. The growth of what is known as open access transmission, whereby larger and larger amounts of energy are whizzed around the grid to meet consumer demand, makes it all the likelier that a sudden and unexpected injection of GIC electrical energy from a solar blast could blow out the system.

"We've been stacking risk multiplier on top of risk multiplier. We've got to preserve our capability!" declares Kappenman, who sees the world's power grids, of which the United States has the largest, as having inadvertently become giant antennas for space weather impacts. Just as a lightning rod provides an attractive, highly conducting alternative to the roof of the house, which lightning might otherwise strike, the power grid, which is designed specifically to be extremely efficient at conducting electricity, serves as a most attractive place for gigavolt space weather blasts to strike. Ground connections at large transformers are the focal points of peril. Much as we in our homes normally ground outlets and appliances, the transformers are grounded for safety reasons, to prevent shorts, shocks, and meltdowns by diverting excess current into the earth. Unfortunately, these same ground connections also serve as portals for electrical shocks coming up from the earth, such as those caused by space weather blasts hitting our planet.

"The scientific community has developed a false sense of security regarding the power industry." Kappenman contends that much of the problem stems back to the space weather event grading system, developed in the years following the Great Magnetic Storm of 1921. It turns out that the classification for the largest space weather storms, a category known as K9, is almost impossibly broad, as though all hurricanes big enough to make it across the Atlantic were classified as Category 5. By that measure there would have been hundreds of Category 5 hurricanes over the years, most of which inflicted little in the way of permanent damage. Thus, the threat of such "Category 5" hurricanes would likely be minimized, since all but a relative few would have been comparatively harmless. Why hasn't the space weather grading system been revised? No reason other than inertia, really, and that the scientific community is reluctant to lose historical data, since the new system, however it was configured, would make comparisons with the "K9 era" more difficult to draw.

Kappenman testified on these matters before the Energy and Environment Subcommittee of the House of Representatives' Committee on Science and Technology on October 30, 2003: "Depending on the morphology of the geomagnetic disturbance, it would be conceivable that a power blackout could readily impact areas and populations larger than those of the August 14, 2003 blackout." That blackout, not space weather related, is believed to have cost between $4 billion and $10 billion in repairs and collateral economic damage. It should be added that in 2003, the global economy was far stronger, and therefore far more capable of absorbing such shocks, than it is today.

AFTERMATH

Poetically, the day Kappenman testified turned out also to be the day of a great solar storm, known in space weather circles as Halloween 2003.

"During breaks in the committee meeting, I was frantically sending out e-mail advisories about the storm," said Kappenman.

Halloween 2003 was much more powerful than the March 1989 wake-up call, but its impact was less severe because it struck mostly at the poles; in the Northern Hemisphere it did not swoop down as far south to the major population centers that consume so much electricity. Nonetheless, Halloween 2003 did cause a brief blackout in Malmö, Sweden. But in the Southern Hemisphere, the damage was far more severe: fourteen or fifteen 400-kilovolt transformers were fried in South Africa, a nation which, in part because of the difficulty in recovering from the Halloween 2003 transformer burnout, has since had enormous problems supplying electricity to its customers, to the point where basic commerce and security have been impaired.

Halloween 2003 has refined our thinking about space weather events, causing scientists to put less emphasis on the overall magnitude of the storm than on where it hits and how fast it moves. Fast-moving CMEs can produce vortices in the Earth's magnetosphere, much the way that heavy wind blowing over the ocean's surface kicks up big waves. The often overlooked magnetic storm of August 1972, not particularly powerful overall, nonetheless created pulsations that, if unluckily directed, had the potential to topple the power grid, according to Kappenman's assessment. It's kind of like when a minor, Category 1 hurricane hits a heavily populated community at high tide—far more devastation than a Category 4 that spins itself out over the ocean.

"This is typical, in that we always seem to find new surprises, usually meaning greater risk, as we look at old storms with modern perspectives applied," says Kappenman, whose sense of urgency regarding space weather threats to the power grid has deepened profoundly since the days of his congressional testimony.

Metatech's work had been partially funded by the Homeland Security Department's Electromagnetic Pulse Commission, which lost its funding in late 2008. Metatech has also suffered in the economic downturn; Kappenman and other senior officials now work as consultants at much reduced compensation, while keeping support staff fully employed. Toward the end of our meeting, I asked Kappenman why he kept on fighting against all economic and bureaucratic odds.

"I would say the odds are against us," he acknowledged. Then he choked up a bit. "It's the social breakdown . . . During Hurricane Andrew, which only affected several counties in Florida, the worst-hit areas, without any electricity or anything, the National Guard, all they could do was leave jugs of fresh water at intersections and hope people would come take them . . . Oil and water pumping would cease, natural gas, too. There would be no ability to refuel a vehicle . . . rail transport. No one keeps fuel at their factories anymore, just-in-time manufacturing took care of that. You can't just restart a nuclear power plant. For one thing, you need for the operators to show up."

FOR WANT OF A NAIL . . .

In his stirring speech to the National Academy of Sciences on April 28, 2009, President Barack Obama passionately declared his dedication to American science, offering the largest federal monetary commitment to basic research in our nation's history. Approximately $47 billion of these stimulus funds are earmarked for energy projects, about half of which will go to repairing and updating the nation's power grid. None of these funds, however, are currently slated to protect the electrical power grid from space weather disruption. In fact, as explored later on, many of the expected expenditures will likely make the power grid even more vulnerable to solar blasts.

AFTERMATH

With the possible exception of those who consider themselves to have been abducted by aliens, space weather has no natural political constituency. As noted, efforts to recruit students and professionals to the field have largely fizzled. Not surprisingly, efforts to promote space weather awareness among the general public have been, to be kind, tame. Earnest scientists bearing PowerPoint presentations just don't seem to catch on to the way the public relations game is played. They have failed to make the threat come alive in the popular imagination. A pity, because if they don't succeed in making their case, a holocaust of cosmic proportions will blow a hole in the human psyche that will take generations to heal. Unless the proper precautions are taken, this megatrauma will indeed happen sooner or later, with the next likeliest time of attack coming with the solar climax of 2012.

Protecting our power grid from space weather assaults is not a new idea. In fact, the electrical power industry has already made some attempts to harden the grid, At several locations, giant capacitors, which store charge temporarily, have been inserted in order to dissipate any electrical overload GICs might cause. However, a Metatech investigation indicates that capacitor mitigation systems cost billions of dollars, and prove effective in only 20 to 30 percent of space weather–related power grid incidents.

A more effective and less expensive mitigation system, Metatech contends, would be to place simple ground resistors that can be used to impede and redirect excess electric current in the time of space weather emergencies. The plan is to place a resistor in the high-voltage transformer's ground connection, which, as noted, is the point at which the power grid is most vulnerable to space weather–induced electrical currents. In a nutshell, CMEs hit the Earth and discharge massive electrical currents into its surface. Ground resistors in turn would protect the transformer from the electrical currents surging up from the ground.

Each resistor would be about the size of a washing machine, and would cost in the neighborhood of $40,000 apiece; with some 5,000 transformers in the North American grid, that works out to $200 million or so, according to Metatech's reckoning. Let's say this estimate is overly optimistic and that significant cost overruns occur, as so often happens in major new programs. Even if the true cost of protecting the power grid from space weather attacks ends up being more in the neighborhood of $500 million, that's about 1/400th of what it cost to bail out AIG for gambling on toxic mortgages, or 1/100th of what Bernie Madoff bilked from his investors. Given that electrical industry revenues in the United States totaled approximately $368.5 billion in 2008, according to the Department of Energy's Energy Information Administration, a one-time space weather security surcharge of about 0.15 percent should amply fund the resistor project. With around 115 million households in the United States, this surcharge would work out to less than $5 per.

This resistor system is believed to have the capability of preventing about 70 to 75 percent of space weather–related power grid incidents were we hit by the equivalent of the great 1921 magnetic storm. It would mean the difference between major inconvenience and societal collapse. In 2008, the ground resistor program was recommended to Congress by the Homeland Security Department's Electromagnetic Pulse Commission, which, as noted, has since lost its funding.

Coming up with the $500 million is not the biggest problem. Congress and/or some beneficent billionaire/foundation could take care of that. The real challenge is political, mustering the will to enact legislation requiring that the geomagnetic resistors be installed. And for that, enthusiastic popular support will probably be required. The utility industry is a patchwork of publicly and privately held entities, rates are largely regulated state by state, and technical specifications are vetted by a variety of different professional organizations. The reason for this mishmash is that the

North American power grid was not constructed as such, but rather is composed of local and regional power systems that have coalesced into a grid over the past century. Only the federal government has the power to ram through red tape. It is a matter of national security. Metatech estimates that the power grid could be largely protected from space weather blasts two years after the resistor development program began. An accelerated program commencing, for example, sometime in 2010 should afford us significant protection by the Mayan end-date of 12/21/12.

The real impediment, one might observe, is the resistor built into the psyche of the electrical utility industry, which spends only between 0.3 and 2 percent of its revenues, depending on the estimate, on research and development. This meager proportion puts it almost dead last compared to other major American industries, less than the pet food industry, according to Wired.com. Computer and pharmaceutical manufacturers reinvest 10 percent or more of their revenues or more in R&D.

"Political gridlock, broken markets, and shortsighted planning have created a slew of bottlenecks . . . Technology alone won't solve this mess, because fixing the grid is not a technology problem—it's a system problem on the broadest scale," writes Chris Anderson regarding the United States utility industry, in Wired.com. Anderson depicts the power grid as an antiquated crazy-quilt of utility fiefdoms, public-private partnerships, and state and interstate regulations and professional standards that somehow functions superbly as a continent-wide unified power distribution system envied worldwide. Bottom line, it is highly reliable but not very efficient, mostly because there is so much redundancy built in. Transmission wires that can carry 500 kilovolts will frequently carry less than half of that, just to be on the safe side in case of surges. Backup generating capacity is considered ample. Fat and happy, the electrical utility industry is thus disinclined to tinker overmuch with the technology that has brought it this far.

Indeed, resistance to the resistors is less about budget than the culture of the utility industry. The industry's objections to implementing a resistor-based space weather mitigation program are more inertial than economic. First, there's the commonsense reluctance to complicate a system that has thus far functioned so well. Inserting ground resistors would probably also require installing high-speed switching circuits to bypass the resistors when necessary, yet another "moving part" that could potentially break down. Plus, the more complex the network, the less control grid operators have over it. Research will indicate whether the best approach is to install resistors as permanent parts of the system or instead to enable power grid operators, upon learning of a potential space weather threat, to activate the ground resistor system.

"We have had no recognition of this potential space weather problem in our power grid network design codes, though we do take into consideration many other environmental factors such as wind, ice, lightning, and seismic disturbances," says Kappenman. He draws an analogy for securing the power grid in this manner to adding seismic retrofits to buildings built before the hazards of earthquakes were fully understood.

True, the addition of ground resistors might marginally impede the flow of current at times, in turn slowing the flow of payments for that current, and also the social and economic activity that the current would have enabled. And malfunctions of these devices, particularly in the early stages, are likely to occur. But the net economic impact would be far less than the $10 billion that it would cost to preventatively shut down the power grid, whether in case of real emergency or false alarm, or the trillions it would cost if the grid were to be shorted out by CMEs. Cataclysm aside for the moment, wouldn't we kick ourselves all the way to hell if the power grid did go down for lack of this simple, affordable, ungrandiose quick-fix?

AFTERMATH

Space weather defense proponents have to build on what we, the public, already know and feel personally. For example, most of us at one point or another have had sunburns; so, too, in a sense, could our infrastructure get sunburned without the proper "SPF." A growing body of research connects solar cycles and CME episodes to everything from increases in heart attacks to stock market crashes. In "Playing the Field: Geomagnetic Storms and International Stock Markets," a 2003 working paper presented to the Federal Reserve Bank of Atlanta, authors Anna Krivelyova of Boston College and Cesare Robotti of the Atlanta Federal Reserve report on a sheaf of studies connecting epidemics of ill health with spikes in geomagnetic activity: "For example, the average number of hospitalized patients with mental and cardiovascular diseases during geomagnetic storms increases approximately two times compared with quiet periods. The frequency of occurrence of myocardial infarction, angina pectoris, violation of cardial rhythm, acute violation of brain blood circulation doubles during storms compared with magnetically quiet periods." The report goes on to state that "at least 75% of geomagnetic storms caused increase in hospitalization of patients with the above-mentioned diseases by 30%–80% at average."

It seems as though a whole new concept of seasonality may be emerging from the study of solar activity. Just as we accept without question that terrestrial seasons influence everything from commerce to poetry to our own bodily rhythms, we are beginning to incorporate the eleven-year cycle of sunspot seasonality into our general understanding of how and why things occur. For example, the evidence presented to the Atlanta Federal Reserve found that stock market indices dropped significantly during and immediately following geomagnetic storms. Investors, unwittingly disturbed by all the space weather titillation, are prone to anxiety and depression, which in turn lead to their making bad decisions in order to get a rush to lift them out of their bad mood, or so the study suggests. (That's all we need: the long-awaited Wall Street recovery done in by sunspots.) It might even prove a boon for astrologers, who will claim that having been born, say, during a solar climax renders one's character more turbulent, or something like that.

The layman's takeaway from the medical evidence presented to the Atlanta Federal Reserve is that your chances of having a heart attack double when the Sun acts up. Although the risks of direct bodily harm from space weather blasts are minuscule compared to the dangers we would face if the power grid went down, the threat of personal injury speaks to people. Scientists seeking to heighten awareness of the dangers that space weather assaults might hold for the grid would do well to capitalize on this personal connection. If the tiny specks of iron floating in your heart can be zapped by space blasts, just imagine what can happen to the electrical power grid, a geomagnetic lightning rod that spans North America. Uncle Sam might get a heart attack! Not scientific, exactly, but it gets the point across. Illustrate with some killer computer graphics of a mega-CME tripping relays and blowing fuses from here to Oshkosh, maybe even throw in a simulation of the user's computer suddenly crashing, and the lightbulbs start going on . . . and then off.

Beyond whizzing off a few e-mails or, better yet, handwritten letters, the format which still counts the most in the formula our representatives use to gauge support for a particular measure, most of us have neither the time nor the inclination to engage in public policy debate. But we all can say our prayers, to whoever or whatever we believe might be helpful, whether it be to our own higher selves or to great God above.

I hope that we get another wake-up call, one larger than the March 1989 event that knocked out power to Quebec, but not very much larger, just enough to scare us into doing what's necessary to spaceblast-proof the power grid from here on in. It would be nice if human nature were such that we didn't need 9/11s to stir us to action—God forbid that the next space weather wake-up call be anywhere near that magnitude—but a space weather blackout that spoiled meat and snarled traffic across America, that just might be a blessing in disguise.

HACK ATTACK

2012, all too often dismissed as an airy New Age doom date, is actually quite a logical time for our enemies to strike. Just as desert heat, blizzards, monsoons, and other components of terrestrial weather have from time immemorial figured into military calculations, space weather must now also be taken into account. As noted in the preceding chapters, the United States is particularly vulnerable, because of our heavy dependence on advanced technologies ranging from supercomputers to satellites to ultrahigh-voltage electrical transformers, all of which can be disrupted by solar outbursts. Our enemies have no doubt recognized this weakness and may well move to capitalize on it during the next violent round of space weather blasts, due in 2012–2013. If they do not strike then, they will have to wait until 2023–2024, the next projected solar climax, which astronomers currently expect to be quieter and lower-energy than the 2012 solar peak. Aside from requiring enormous patience, not a trait for which villainous predators are normally known, waiting until 2023–2024 would also mean gambling that our power grid and other vital infrastructure had not been protected, or "hardened," by then. Why pass on what might be their last, best opportunity for decades to come? Moreover, waiting for the 2023–2024 solar climax might well mean missing the opportunity to take advantage of the once-in-a-century global economic crisis that no doubt will still be taking its toll in terms of deferred maintenance, debt burden, and distracted focus in 2012.

How might an enemy take military advantage of a space weather assault? Let's say, God forbid, that a group of hackers, whether they be al-Qaeda operatives, anarchists bent on disruption and extortion, or coordinated, high-level Russian or Chinese cyberassault teams, have prepared an attack on the electrical power grid of North America. During the forty-five minutes or so from the time the solar blast was sighted by the ACE space weather satellite (assuming that the aging, almost fuel-less sentinel has not sputtered out by then), power grid operators would be frantically shifting power loads, disconnecting transformers, starting up capacitors, and otherwise stressing the diagnostic and operational capabilities grid. That's precisely the time for the hackers to strike, when grid operators have far less time, attention, and resources to repel the computer assaults. What might otherwise have been contained as no more than a pesky hack attack could, when amplified by space weather pummeling, turn out to be a megadisaster of the scale that the National Academy of Sciences predicts.

"Electricity Grid in U.S. Penetrated by Spies," a recent Wall Street Journal online article (April 8, 2009), claims that the hacker assault on the power grid is already under way, though few facts are offered to support that frightening claim. Homeland Security Director Janet Napolitano declined to confirm or deny the report. White House spokesman Nick Shapiro was similarly noncommittal. CIA analyst Tom Donahue was quoted as saying that such assaults have already taken place elsewhere, though he offered no specifics.

"We have information, from multiple regions outside the United States, of cyber intrusions into utilities, followed by extortion demands," claimed Donahue.

Unnamed sources raised the specter of Russian and Chinese government operatives burrowing into our grid.

"They [the Chinese] are all over the place. They're getting into our university systems, contractor systems, hacking government systems. There's no reason to think that the electrical system would be immune as well," said the anonymous government official.

The *Journal*'s piece is light on facts and heavy on "yellow peril" and seems more like a public relations plant designed to drum up support for funding cybersecurity than the solid journalism for which that august publication had formerly been known.

"I am almost beginning to believe that every time the U.S. government wants to make cybersecurity a big issue, the intelligence agencies oblige by releasing a report on unknown persons or organizations hacking into the power system computers," says Bruce Wollenberg, professor of electrical and computer engineering at the University of Minnesota. Wollenberg, who has for years investigated reports of computer hackers invading the power grid, observes that "just because someone figures out how to get inside the bank at night doesn't given them the combination to the safe."

Hackers would have to be extremely knowledgeable about power grid operations to do more than shut down a few computers, an act which in and of itself would cause significant inconvenience, but probably no serious outages, according to Wollenberg.

There is no doubt, however, that a team composed of knowledgeable hackers and power grid operators could do serious damage to the grid, perhaps even shut it down in part or whole. In 2007 a simulation performed at Idaho National Laboratories demonstrated that a grid operator enlisted for the exercise was able to penetrate the system's defenses and damage a generator by tripping a circuit breaker, thus causing a shutdown of a small section of the simulated power grid.

OUTSMARTING THE SMART GRID

Since energy efficiency is tremendously important these days given the high economic and ecological costs of waste, tremendous emphasis is being placed on "smart grid" upgrades, which have been heavily funded by the Obama administration's stimulus program. The goal of these upgrades is to increase the efficiency of power transmission and consumption by making fuller use of the grid's current capabilities, meaning less backup and redundancy, less margin for error. Unfortunately, smart grids can also be more easily outsmarted than conventional grids, which really don't do much thinking at all.

One method of improving grid efficiency is to introduce wireless "smart meters." It's kind of like replacing traffic lights with (computerized) traffic cops who are better able to address specific situations and therefore more quickly unsnarl jams. The downside of smart meters is that they can be compromised by hackers who trick them into indicating that which is not so. For example, a nefarious hacker could instruct all smart meters on a section of the grid to request more power. Too many such requests occurring simultaneously could lead to the overload and shutdown of a power plant, soon resulting in a cascade of shutdowns around the grid.

The new smart grid will be far better at taking commands, and therefore far more susceptible to commands made by impostors. False "shut down" alarms can cause mischief, while false "all clear" messages, issued when there is in fact a problem on the grid, such as might be caused by a space weather blast, could prove phenomenally problematic. Of course, smart-grid designers are building in numerous checks and balances to keep hackers from crippling the nation's power supply. One tactic is to make sure that the smart meters in any given section of the grid are divided into groups that use different communications protocols, meaning that a successful hack attack could fool only a portion of the meters, perhaps causing trouble but not catastrophe.

According to Datamonitor, a London-based industry analyst, smart-meter penetration in the United States electrical power grid should reach 89 percent by 2012. Smartening up the grid, thus improving energy efficiency and saving perhaps hundreds of billions of dollars, would be a solid accomplishment for President Obama and the Democratic Party to point to as they seek to maintain control in that presidential election year. But at this writing, given the absence of support for hardening the grid against solar blasts, there seem to have been no provisions made for the possibility of hackers coordinating attacks on the grid with space weather blasts, thereby amplifying minor solar assaults into major disruptions. All the hackers would have to do to is sync in the ACE real-time solar data page on NOAA's Space Weather Prediction Center's website, and program their own computers to sound an alert when certain key space weather variables surpass critical thresholds, thus triggering the initiation phases of the hackers' own assault program. Beyond this it is wise not to elaborate. No doubt the cleverest assault teams are already way on top of this, but why toss a bone, say, to some dim-witted but technically competent al-Qaeda wacko waiting to detonate that one big blast that will rocket him to Heaven and the rest of us sinners to Hell?

Let's give our leaders something they can really brag about. As long as they are planning to revamp the grid anyway, tell them to take the extra step and include protection against the space weather blasts that either alone or in concert with enemy attack could bring our nation down, and with it much of the world.

DISCONNECTED

"MAKE ME LAUGH!"

The telephone rang in the middle of the night and that was the command. Liz, my girlfriend at the time, winter of 1988, was driving back to her home in Tumba, Sweden, not far from Stockholm, and had gotten caught in a snowstorm. Frightened that she would crash or get marooned on the highway, she had called from her car to my studio apartment in New York so that I would calm her down, steady her hand. In my 4 a.m. confusion the only joke I could think of was a giraffe imitation, a sight gag that doesn't work over the phone, but I babbled on gamely for a half hour or so until Liz, laughing in spite of herself, pulled safely into her driveway.

Precious and important moments between two lovers made possible by modern technology, but which signals were bouncing off what and where so that we could connect? Liz's mobile phone emitted a binary stream of 0s and 1s that connected with various reception towers as she drove down the highway, each tower in turn amplifying this transmission and then relaying it up to a telecommunications satellite, probably a LEO (low Earth orbit) satellite, circling the North Pole at a relatively low altitude in the range of 400 to 1,000 kilometers (250 to 600 miles).

AFTERMATH

On the way back to Earth, our conversation would have beamed down through the exosphere, past any manned spacecraft such as the space shuttle, which usually orbits at around 300 kilometers (185 miles), on down through the stratosphere, 51 kilometers (32 miles), through the tropopause, a boundary layer at about 7 to 17 kilometers (4.5 to 11 miles), and on into the troposphere, where we all live.

There are usually about two hundred LEO satellites plying the atmosphere over the North Pole. Their proximity to Earth enables phone transmissions to travel back and forth to the surface with very little delay, on the order of 20 to 30 milliseconds, meaning few annoying lags or echoes in the conversation. LEOs require less signal strength than satellites that must transmit from higher orbit. However, their low altitude does make them more susceptible to the Earth's gravitational pull, which increases, as Newton discovered, in inverse relationship to the square of the distance an object is from the center of the Earth. The lower a satellite's altitude, therefore, the harder it must work, and the faster it must travel to avoid crashing to the ground. Orbiting the pole is kind of like orbiting a mountain; the satellites duck in and out of transmission range rather frequently. To compensate for this difficulty, LEOs operate in constellations—as one dips below the horizon, the signal it had been receiving is simply kicked over to another LEO still within sight.

During the late 1980s, there was a lot more at stake regarding the LEO satellite system than just keeping callers like Liz and me connected. It was the last years of the Cold War, and itchy trigger fingers on both sides of the divide depended on spy satellites, which also tend to ply low orbits, to provide the stream of reassurance needed to keep the nuclear war button from being pushed. HEO (highly elliptical orbit) satellites trace what are known as Molniya orbits, elliptical paths originally calculated by Soviet military experts to enable satellites to cruise very slowly, in fact, to almost hover above the North Pole at very low altitude, 300 kilometers (185 miles). Spy satellites for both the United States and Russia plied and continue to ply HEO orbits, peeking down upon each other's territory from their low-orbit vantage points at the North Pole. In the simplest terms, these spy satellites, also known as keyhole, or KH-class, satellites, operate much like the satellite that carries the Hubble space telescope, except that in this case the telescope and/or listening device points down at the ground.

What if a CME solar blast disrupted military satellite operations, now or, worse, back in the ultratense Cold War era? Not being able to count on its space resources makes the United States Air Force a bit jumpy:

"For example, was a solar radio burst or a thunderstorm the cause of a communications problem, or was it caused by someone trying to deny the use of the communication band?" says Major Herbert Keyser of the U.S. Air Force, Space and Intel Weather Exploration.

Normally, the Earth's magnetic field protects everything from satellites to skin by forming a barrier in space that deflects incoming CMEs into the Van Allen radiation belts orbiting high above the planet's surface. Whenever there is a crack in the Earth's magnetic shield, such as the immense pole-to-equator hole recently discovered to have opened up, satellites passing through that crack become vulnerable to the errant solar outbursts. Just how many satellites have already been lost to solar radiation is very hard to determine, because nations and corporations like to keep this information quiet; not good for general morale. We do know that upon passing through the South Atlantic anomaly, a California-sized crack that opens periodically in the Earth's magnetic field, several satellites were rendered nonoperational, including, ironically, a Danish research satellite sent up to study the crack. In January 1994, two Canadian telecommunications satellites suffered power outages during a period of enhanced energetic electron fluxes at geosynchronous orbits, disrupting communications nationwide. The first satellite recovered in a few hours; the second satellite took six months to restart and cost $50 million to $70 million.

The good news is that satellites are much less vulnerable than the power grid to solar blasts, for the very good reason that most of them have been specifically hardened or protected against such assaults. However, only military satellites are thought to be sufficiently hardened—think bulletproof glass—from the megastorms that could climax in 2012. What's more, budget constraints and other pressures have forced the military to shunt many of its orbital operations onto comparatively defenseless commercial satellites. On balance, however, the satellite system that we have inherited from the Cold War era and thereafter is well protected, perhaps even over-protected, from space weather assaults.

"Radiation belt models are overly pessimistic about the amount of degradation that will occur [from solar blasts] and have led to the costly overdesign of satellites in some orbits," says David Chenette, of Lockheed Martin Space Systems Company. Chenette explains that a satellite is hardened against solar blasts by its susceptible circuitry being encased in dense, heavyweight shielding. Each extra pound ups a satellite's launch cost by about $40,000.

Ironically, the greatest space weather threat to the satellite system comes via the electrical power grid that supports it from the ground. Can't sustain the flow of telecommunications, military surveillance, or any other sort of satellite data if the ground receivers aren't plugged in. So the best way to protect our satellites from space weather blasts is to do our best to ensure that ground support remains up and running.

12/21/2012 = 12/7/1941?

Fear and loathing aside, it's a pity that former Secretary of Defense Donald Rumsfeld, coiner of the almost mystical phrase "unknown unknowns," has dropped out of sight. Not, Lord knows, for his perfidious prosecution of the war in Iraq, but rather, as touched on in my previous book, because Rumsfeld had the foresight in early 2001 to warn of a "Space Pearl Harbor" attack, declaring his commitment to protect the United States from such a calamity. Had we not subsequently suffered the terrible misfortune of the 9/11 terrorist attacks, which led, however circuitously, to the quagmire in Iraq, Rumsfeld might not have abandoned his commitment to space defense and led us to the level of security that we need and deserve. Both the youngest man ever to serve as secretary of defense (from 1975 to 1977 under Gerald Ford) and also the oldest, under the junior Bush, Rumsfeld has an invaluable long-term perspective. In Rumsfeld's scenario, the remote American outpost that would come under attack is not a Hawaiian harbor full of U.S. warships, but an orbit full of our military, navigational, and telecom satellites, particularly those in the lower orbits most accessible to ballistic missile attack.

The weapons are already in place, various arsenals of ASATs (anti-satellite weapons) held over from the Cold War competition between the United States and the Soviet Union to dominate space. In September 1985, the United States Air Force successfully tested an ASAT shot off the back of a high-flying F-15 fighter jet. The ASAT did not explode but rather crashed head-on into its target, an aging LEO satellite known as Solwind, orbiting at 555 kilometers (345 miles). The American ASAT is considered quicker and more agile than the ASAT first developed by the USSR in the late 1960s, a ground-based ballistic missile that had to be fired only when the target LEO satellite flew overhead. The Russian

ASAT would then track the LEO for an orbit or two, move closer, and then explode into it, the satellite equivalent of a suicide bomber. (In 1987, the Soviets grabbed clumsily for space dominance by launching a prototype for a "space battle station," but it never reached orbit and fell into the Pacific Ocean.) In 2007, China became the third nation successfully to test an ASAT, a missile system that combines American and Russian features. Like the Russian ASAT, the Chinese version was launched from the ground, but like the American version, it utilized what is known as a "kinetic kill vehicle," which does not explode but rather crashes into its target.

Beneath the three space superpowers, there is now a second echelon of about a dozen nations and several private entities able to take out LEOs, albeit indiscriminately. The crudest ASAT is simply a primitive rocket launched into low Earth orbit and then detonated, creating debris that would incapacitate whatever else happened to be in the same orbital path. The attacker would therefore risk losing some of his own satellites, and his allies'. Depending on the manner and, particularly, the specific altitude at which the assault is made, the space junk that results from destroying a satellite militarily can pose a hazard for scores of other satellites, much as blowing up an automobile in the middle of a crowded interstate highway would imperil many other vehicles.

Could a rogue state, terrorist group, and/or evil billionaire trash the LEO satellite system and get away with it? Probably not, since such an attack would require constructing and arming a large rocket, and then either building or obtaining a satellite launch facility, of which there are believed to be two to three dozen around the world. Spy satellites would certainly spot from orbit any unauthorized antisatellite weapons system large enough to be effective. Whether and how fast the weapons discovered would be destroyed by military action is of course another question, as is whether an antisatellite installation could be assembled quickly enough to be fired before reprisals came. Most likely, a rogue attack on the LEO system would be suicidal, incurring immediate, lethal reprisal from the United States military and the global coalition it leads. But by then, the terrorists would already have done their damage.

What terrorists and rogues just might be able to do, however, is to trick one of the superpowers into thinking that a nuclear attack or some other space calamity is about to occur. Although Ronald Reagan is about the last person one would think of to ever have supported terrorism, his 1983 Star Wars initiative did yield some unexpected and unsettling results. MIRACL, a megawatt-class chemical laser ASAT based at the White Sands Missile Range in New Mexico, has had its problems, and its funding has been largely withdrawn. However, one surprise discovery made while testing MIRACL remains of particular concern.

"A low-power [30-watt] laser intended for alignment of the [MIRACL laser ASAT] system and the tracking of the satellite was the primary laser during the test, and it appeared that this lower-power laser was sufficiently powerful itself to blind the satellite temporarily, although it could not destroy the sensor," according to an analysis conducted by the Union of Concerned Scientists.

The discovery that a small, low-powered laser bought from an electronic supply house or over the Internet and easily concealed, only 1.5 meters in length, could actually disrupt a satellite's functioning was a shock. What havoc could be wreaked if, say, fifty or a hundred of those lasers were shot simultaneously at key satellites? What if the attack laser's power were doubled or tripled, calibers that are still commercially obtainable? Reports that one such laser fired from somewhere in China "lit up" a United States military satellite in 2006 are credible, though unconfirmed. It is not the portable lasers' destructive capability so much as their ability to interfere with the military surveillance data stream that causes anxiety. Suppose that just as North Korea were readying another ballistic missile test, American spy satellites lost the ability to track the weapon's whereabouts. Tokyo? Alaska? That's the kind of gut-level decision-making that could lead to an unfortunate nuclear exchange, particularly, one might imagine, in the raw-nerve geopolitics we may well be looking at in 2012.

The space weather tumult of 2012 will provide cover for enemies seeking to disrupt the satellite system. What better time than a solar climax to launch a military attack on the LEO satellite network, already under siege from the heavens? Nature will serve as the evildoer's unwitting ally, making it that much easier to inflict maximum damage on our orbital networks and on the electrical grid that supports it from the Earth's surface. A serious breakdown of the satellite telecommunications, even if only temporary, could well produce a domino effect, triggering anarchic and predatory behavior. Criminal organizations would ruthlessly exploit gaps in surveillance as, no doubt, would their comrades, the terrorists. Lack of electronic oversight for financial transactions would let a thousand Bernie Madoffs bloom.

Disabling the satellite network, in short, may be the first step to a global coup d'état. The first nation or entity to restore this network will control military and business communications and will, quite literally, rule the post-2012 world.

"The consequences of war in space are in fact so cataclysmic that arms control advocates ... would like simply to prohibit the use of weapons beyond the earth's atmosphere ... The United States has become so dependent on space that it has become the country's Achilles' heel ... It's only a slight exaggeration to say that an M1-A1 tank couldn't drive around the block in Iraq without them [satellites]," writes Steven Lee Myers, in *The New York Times* (March 9, 2008). Myers ticks off horrendous eventualities, such as the failure of the global economic system, air travel, and telecommunications. "Your cell phone wouldn't work. Nor would your A.T.M. and that dashboard navigational gizmo you got for Christmas. And preventing an accidental nuclear exchange would become much more difficult," observes Myers.

Civilization's ability to survive such an assault depends in no small part on how quickly damaged satellite operations can be shifted to higher orbit. ASATs are geared for use against low-Earth-orbit satellites only and cannot readily be adapted to attack higher orbits. If the LEO satellite system were disrupted, basic military and telecommunications services would, in all likelihood, be rerouted to satellites at higher altitudes (assuming, of course, that the command and control required to facilitate this transfer had not been knocked out on the ground). In essence, each orbital level—low, medium, and geostationary—operates as its own separate satellite network, with virtually no communication among those levels, except as might occur via the ground. There is, however, an implicit redundancy built into the overall system, since satellites operating at one orbit can, when the need arises, take over many of the vital functions that would otherwise have been handled by craft at other orbital levels. It's not ideal, perhaps somewhat like trading an SUV for a golf cart, but any vehicle is better than none.

In the event of system-wide incapacitation, the bulk of the LEO satellite data load would likely be shifted 20,000 miles or more higher to the network of GEO (geosynchronous, or geostationary, Earth orbit) satellites, so named because GEOs provide a steady target, always remaining exactly 35,786 kilometers (22,090 miles) above a given spot on the equator. (If a GEO were fixed above a point that was not on the equator, it would appear to move in a north-south direction and would therefore be harder to target.) The fact that GEOs are stationary enables them to track moving targets ranging from weather systems, such as the snowstorm Liz was caught in, to airliners to cruise ships. A major advantage of GEOs, therefore, is that they can be targeted by fixed antennas, which are a lot less expensive and more reliable than tracking antennas that constantly adjust to the movements of their targets. Thus, the satellite dish on the roof of your home can remain in one position, pointed toward its GEO high above the equator, rather than needing to automatically shift position to keep on receiving the data stream.

Because it is so high up, a GEO satellite commands a superior view and can communicate with up to 40 percent of the surface of the Earth; three GEOs can with a little overlapping pretty well cover the planet. The extreme altitude also gives GEOs the advantage of longer life span: they are beyond the range of attacking missiles and even immune to corroding friction, the atmosphere way up high being so thin. GEOs are less well-suited for applications such as telephony that require instantaneous feedback between ground station and satellite, their extreme altitude causing an unavoidable lag time of around 125 milliseconds that causes annoying pauses and overlaps in conversations. Computer programs have been developed to adjust for this, delaying a response transmission until the previous one has been completely received. Although the results are still not as good as with conversations conducted with LEOs, which operate with only a few milliseconds of delay, GEO telephony will do just fine if our LEOs are pinched.

Hijackers might daydream about seizing the Boeing Sea Launch, a U.S.-Russian commercial operation out of Long Beach, California. Sea Launch is an oceangoing launch site in which GEO satellites are loaded and assembled on a ship, and where the launch rocket is laid horizontally on a self-propelled platform, about two football fields in size, that used to be an oil rig in the North Sea. Ship and platform sail together to Kirimati, an island on the equator in the Pacific Ocean, where the satellite is loaded onto the rocket, which is then erected, fueled, and launched into geostationary orbit. The Sea Launch is one of the most highly secured and monitored craft in the world, meaning, fortunately, that seizing it for terrorist purposes would make a much better Hollywood thriller than a real-world military assault plan.

Not to say that GEOs are completely invulnerable. One amusing and frightening story about GEOs is that they are routinely hacked by electricians in the Amazon rain forest, where cell phone towers are hard to come by. According to Wired.com ("The Great Brazilian Sat-Hack Crackdown," April 20, 2009), locals who wish to communicate telephonically with each other when deep in the jungle have learned to hack into the GEOs and chat with each other on Uncle Sam, or Uncle Telco, doesn't really matter which. No malice is intended by these hackers, just theft, say, of the opportunity to tell their family whether or not they will be at home that night, or maybe to brag to a buddy about discovering which beer makes you burp the loudest.

Had GPS (global positioning system) been up and running back then, it might have helped Liz find her way through the snowstorm; it certainly would have done so for marine vessels and aircraft caught in the same blizzard. GPS NAVSTAR, which became fully operational in 1993, operates as a constellation of twenty-four to thirty-two MEO (middle Earth orbit) satellites swarming around the planet at 20,200 kilometers (12,500 miles) in altitude. GPS satellites take anywhere from two to twelve hours to complete an orbit, since some swing wide and others carve tight circles, to cover all the angles required to provide precise coordinates for any spot on Earth. The GPS, though in worldwide commercial use, is and will always remain a military construct. Destroying the GPS satellite system would be next to impossible, except for an all-out attack by Russia or China. Hackers try periodically, but they barely make a dent. In 2003, devotees of Phrack.com, the now defunct computer hackers' website, did acquire the ability to jam some GPS ground receivers, but nothing ever came of it and no technological breakthroughs appear to have been achieved.

AFTERMATH

With all the forces conspiring to disconnect us—hackers, enemies, rogues, terrorists, CMEs—it does seem likely that it will happen one day soon. The horrors and dangers of such a catastrophe, touched on above, are easy to imagine. Less obvious are the benefits.

What's the longest you've ever gone without access to electronic communication? No phone, no fax, no Internet, no television, nothing with an electrical pulse? It's make peace with yourself and your circumstances or die, a fact I can vouch for, having spent two four-month periods in the late 1980s completely disconnected while residing at the Fondation Karolyi, an artistic foundation in southern France. The helplessness is the hardest part. If an emergency arises anywhere except where you happen to be at the moment, you're not going to be able to help. It's not an option to be so disconnected if one has dependents. Or if one seeks opportunity, which might well pass one by in favor of someone else who can be rung up conveniently. Members of the military are forced to get used to having limited telecom access, albeit within a highly structured support system that compensates for this deficiency. For most of the rest of us, though, navigating life without telecom access is kind of like learning how to live without crossing the street—you can do it, but it's not convenient.

But once you get used to it, being disconnected electronically is really quite peaceful. Losing telecom is like losing one of your external senses while gaining, perhaps, an internal one. Less external communication leads to more internal communication. There is a steady, soothing improvement in one's mental acoustics. Going without a telephone is like going from marionette to independent person, freed of the strings, in this case, telephone calls, that jerk one about throughout the day. Not to argue that the state of disconnection is in any way superior to being plugged in, but rather simply to suggest that the contemporary, data-rich lifestyle is just too much of a good thing. Lord knows we shouldn't need a calamitous collapse of the satellite system to relearn the simple pleasures of peace, quiet, and face-to-face communication, and that boredom is not the lack of stimulation, but rather the lack of appreciation of the moment. A couple cell-phone-less weekend retreats listening to the birds tweet could teach us that. But in Mayan terms, rediscovering ourselves and our centers in the quietude of telecom disconnection would be the silver lining of the 2012 cloud. It would indeed be the birth of a new era where solitude and connectedness are treasured accordingly, where the tender needs of the soul are balanced with the enthusiasm of the mind.

THE ERA OF WHEN

Much has been written about how militant Islam has replaced communism as the enemy of the West. That's true enough, as far as it goes, though it seems more accurate to say that the scope of our fears has widened to include climate change, terrorism, economic collapse, and now, perhaps, threats from the Sun. But the real change is that our enemy is no longer a who or a what, but a when.

Twenty-twelve has caught fire in the popular imagination precisely because it is a date. As such, it taps into our budding awareness that the hopelessly complex, utterly contingent global socioeconomic system that we have constructed with speed-of-light information technologies is eventually going to glitch big-time. Whether brought down by warfare, natural catastrophe, terrorism/anarchy, greed, whopping technological mistakes, or some dominolike combination thereof, it's only natural for such a correction to occur. Question is, how big a correction are we talking here? World War III big or just global recession big?

The 2012 era is a global juggling act, with the balls, bowling pins, and chainsaws whirling literally all the way up through the atmosphere and beyond. Like all juggling acts, it has to end. What goes up must come down, perhaps gently, perhaps crashing on our heads. Focusing on future doom dates, the latest one being 2012, is therefore neither foolish nor gloomy, as oh so many have charged. Rather, it is humble and realistic. Humanity is in the early stages of unprecedented global collaboration and interdependency. We can't expect to perform flawlessly the first time. Mistakes of unprecedented magnitude will be made. Some can and will be learned from, but others just have to be avoided. It is urgent that we identify the most dangerous areas of vulnerability, such as the North American electrical power grid, and also the satellite telecommunications system, that keep our global juggling act afloat.

Back when Liz and I were dating, no one gave much thought to 2012. The notion of apocalypse, however, hung like a mushroom cloud. The sense that civilization was always only an hour or two away from nuclear annihilation was just part of the ambience of the Cold War, a nervous tic in everyone's psyche. There was nothing divine or supernatural about it, no coalescence of uncanny coincidences that there seems to be today—just the megatons of nuclear weapons and the continent-sized waves of killer radiation that their detonation would release. In a way it was oddly empowering, in that we human beings, not God, not nature, controlled our own destiny. Plus, it had to be admitted that the MAD (mutual assured destruction) doctrine worked pretty well: neither the United States nor the Soviet Union were anywhere near capable of absorbing the retaliatory blow that the other would dish out if sneak-attacked, so, unless a madman rose to power in either nation, odds were always good that the Button would remain unpushed. Although we remain in a MAD situation today, with Russia and the United States each possessing more than enough nuclear weapons to do the other's society in, the Cold War is over. Nuclear war was averted, and though still a distinct possibility, it no longer weighs on us the way it once did. However, that doesn't mean we're in the clear.

China is the obvious choice to replace Russia as our next superpower enemy, but skillful diplomacy from the days of Richard Nixon and Chou En-lai to the present have kept that relationship, prickly at times, from turning bare-knuckles adversarial. However, the will to dominate is not always mastered by diplomacy. Of greater political significance in the long run is the 2008 Chinese manned-mission success, coupled with their declared intention to colonize the Moon. They who mine the Moon will rule the Earth in the second half of the twenty-first century. They will control not only the Moon's mineral supply, but also an immense low-gravity environment in which to test and create advanced alloys and other materials superbly suited for military applications in space. In addition to being the inspirational leaders of the greatest new stage in human evolution, they will have the perfect opportunity to take military control of the planet's telecommunications and space defenses as they return to Earth orbit.

Perhaps the Chinese are destined to control the Moon, impelled by the twenty-second-century dream of exporting some of their massive population up there. Comparatively poor in natural resources, China may also see energy freedom in the form of a rare isotope of helium gas, known as helium-3, abundant on the Moon but extremely scarce on the Earth. As explored in my previous book, helium-3 may well turn out to be the most powerful, and cleanest, fuel for controlled nuclear fusion. The comparison often cited is that a space shuttle cargo load of helium-3 could meet the United States' electrical requirements for one full year. Helium-3's most ardent exponent is Harrison "Jack" Schmitt, well known as a member of the Apollo 17 crew in 1973, the last manned mission to the Moon. The only geologist ever sent to our natural satellite, Schmitt discovered helium-3 there, and quickly grasped the significance of the vast lunar reserves. Being so light, the gas bounces off the Earth's atmosphere, but because the Moon has almost no atmosphere, helium-3 flies right into the Moon's rocky, sandy surface, known as the regolith, and stays there, loosely bound up. So whoever figures out how to "sift" the Moon for helium-3 could be worth about ten quadrillion, give or take. The Persian Gulf of the twenty-first century, the Moon may well become.

Since returning to New Mexico to serve a term in the United States Senate from 1976 to 1982, Schmitt has devoted much of his time to campaigning for a lunar mission to mine the Moon for helium-3. He was so dedicated to that mission, some say obsessed, that New Mexico voters found him lacking in local bread-and-butter issues and rejected his bid for reelection. However, Schmitt's great lunar vision does not seem to have been lost on the Chinese.

Immortal sagas are born of eras such as the one we are now entering. Healthy, searching souls voyaging into space to fulfill the radiant Mayan promise of the new era of enlightenment that begins on 12/21/12. Humanity, under whatever set of flags and banners, bringing life to the heavens, colonizing lands which have no natives to be slaughtered in exchange for wealth—what a divine opportunity for bloodlust to begin devolving as a human trait!

Aftermath Scenario: 2012 Begins

It was as though we had all been banished to live on balconies and rooftops without any guardrails. That weird, shaky tingling you may get in your feet while looking down from great heights was how it felt to walk around everywhere, all the time.

The beginning of 2012, not the technical calendar beginning but rather the dawn of the general sense that the prophesied changes were now under way, came relatively late in the year, during the first few weeks of September 2012. September had always felt more like a new year anyway: end of summer, new school year, Jewish New Year, back to work, autumn begins. Ever since the new millennium started, the biggest catastrophes to hit the West have come during this changeover time: the credit-crunch depression that began with the stock market crash of September 17, 2008; Hurricane Katrina hitting New Orleans on August 29, 2005, and the weeks of chaos that followed; and, of course, September 11, 2001.

Right after Labor Day 2012, the sun storms that had been on the rise all year long spiked viciously, belching out billion-ton X-class blasts of solar radiation that tortured the world's power grids, causing blackouts and brownouts throughout the Northern Hemisphere. Ecologically efficient "smart grids," which steer electricity demand from mainstream generators to renewable energy suppliers, proved particularly susceptible to these blasts because of their delicate computer circuitry; much of the $25 billion allocated by the Obama administration to fund such grids went up in smoke.

The network news executives couldn't have been happier. Blackouts are fun to cover, providing that rare mix of people pulling together in time of adversity, directing traffic when streetlights go blank, that sort of thing, and just enough looting and dark-of-night mayhem to make it honest news. Cityscapes gone dark, red-faced officials squirming in the spotlight, toilets not flushing because the water pumps that supply them had no electricity, eco-friendly features touting the virtues of going "off-grid," all this plus a treasure trove of file footage from previous blackouts, starting with the last major blackout caused by solar blasts, when Hydro-Québec of Canada went down in November 1989, depriving several million users of electricity for about eight hours. Important, entertaining, yet not so urgent that the commercials have to be pulled. All in all, a good week of news.

Everything was expected to return to normal in a few days, just as in March 1989. But as the sun storms and blackouts of September 2012 rolled into a second week, and then a third, repair crews became overwhelmed. Reports that more than 100 transformers had been blown across North America and now the European Union were vigorously denied. Loopholes in the contracts of the now privatized power utilities allowed them to divert, at a hefty premium, electricity from inner cities to affluent neighborhoods, sparking protests and riots which the police, fire, and other emergency services could not contain.

Our sophisticated technological infrastructure balked at the power shortage like a baby yanked from the breast. Basic telephone communication, cell and landline, went in a few weeks from taken for granted to an exciting surprise, like the way getting a telegram delivered to your door used to be. Internet service proved surprisingly hardy, almost heroic in the way it would flicker back on at odd moments. News reports that the satellite system was also under serious siege by the solar blasts would have caused more outrage and alarm, that is, if more people were able to tune into the news.

Computer hackers, like jackals, raided sensitive databases whenever they came back online. A weird manifesto, *Phrack 'Em All*, began to make the rounds. It declared that the world had entered a new and peculiar era of warfare, wherein major natural disasters would be seized upon as opportunities for military assaults. The perpetrators were a loose amalgam of establishment enemies known as Phrack, a name taken from Phrack.com, which had once been the cyberworld's leading computer hackers' website. Over the years, Phrack.com had published detailed instructional manuals for all sorts of disruptive devices. In 2003, users were shown how to construct GPS jamming devices from inexpensive components easily obtained from local electronics supply shops. Most of Phrack's pranks were pretty sophomoric, except for those intended to hack into information data banks and steal consumers' identities. But the movement that took the Phrack name was no joke at all.

Whenever an earthquake struck, a volcano erupted, or a major hurricane made landfall, an amalgam of disruptors would take advantage of the chaos and strike wherever the government or international organization in charge of responding to the natural catastrophe was deemed weakest. Attacks were never to be launched at the catastrophe sites per se. For example, had Phrack struck during the great Chinese earthquake of 2008, the stricken Chinese province of Sichuan would not have been hit. Instead, the Phrack attack, as the media took to calling such episodes, might have been on Chinese cargo ships crossing the Pacific, or on the Hong Kong stock market, or on the frantic construction effort in the run-up to the Summer Olympic Games. Politics didn't matter to Phrack: Israel and Syria were equally fair game for attack, likewise India and Pakistan, Japan and North Korea. Of course, the United States and Western Europe, having the greatest range of responsibilities, were hit most frequently, and the hardest.

Like the Vandals and the Visigoths who brought down the Roman Empire, the Phrack movement had one overriding principle: disruption. Phrack struck when Nature struck, everyone knew it, and the only question was where and how hard. Like al-Qaeda, this enemy was loosely organized; it had little discernible hierarchy and no central source of funding. It was simply a consensus understanding that civilization's power structure needed to be brought down, and the way to accomplish this was to inflict damage during the chaos of natural catastrophes. The only ideology that could be inferred from this blatant, bloody opportunism was that Nature was an insurgent force implacably opposed to the tyranny of civilization, and that Phrack's warriors were simply coming to Her assistance. Some even believed that the mounting wave of catastrophes was proof positive that Nature was trying to throw off the yoke of human oppression. But imputing ideology to this movement was like anthropomorphizing entropy—attributing human characteristics to a phenomenon of nature may help one think about it, but the conclusions one draws will probably be wrong.

For exponents of the New Catastrophism, the emerging philosophical school that would dominate the first half of the twenty-first century, the Phrack movement set visions of doctoral dissertations dancing in their heads. The simplest way to disrupt global civilization and throw it into a state of anarchy is to destroy the electrical power grid that holds it together. Fortunately, no one in their right mind would want to disrupt the grid, if for no other reason than that any such perpetrator, particularly any independent actor, would more than likely be caught and suffer a swift and fiery death. Unfortunately, as suicide bombers have demonstrated, there is an increasing willingness on the part of malefactors to sacrifice their own lives to obtain their goals. For the dedicated intellectual, this is an irresistible conundrum. For the rest of us, it's ulcers.

By the end of October 2012, few actual Phrack attacks had occurred, in part because, like the "revolutionaries" of the 1960s, the Phrack movement was mostly just bluster, at least for the moment. Plus, there was the fact that Phrack attacks could not take place when the power grid was down, which had been the case more often than not for almost a month. A constructive analogy was drawn between the Phrack insurgency of 2012 and the Extra-Strength Tylenol tragedy of 1982, a deranged episode in American history wherein bottles of the medicine were poisoned and then set back on the shelves, ultimately killing several random victims. When Halloween came later that year, there was a bizarre uptick in malicious pranks played on trick-or-treaters, such as poison put into candy and razor blades slipped into apples. Horrible as this was, our worst fears, that American society had now settled permanently into this depraved state, and that a happy Halloween, and along with it basic trust in one's neighbors not to harm innocent kids, was a thing of the past. But the dementia passed, and in two or three years, Halloween was back bigger, better, and even happier than ever.

By Halloween 2012, there was a flickering feeling that we had entered a new and wildly unstable era, that at any moment our lives could be unplugged, by solar blasts, by anarchist hackers, by politicians' corruption, and by bankers' greed. Much as during the stock market crash/credit crunch that began in September 2008, there festered a dark suspicion that conditions were not temporary but, rather, a long-term downshifting to a lower level of prosperity and security.

This sense of foreboding deepened into full-on dread when a peculiar meteorological formation, a cluster of hurricanes, emerged just off the West African coast several days into November, right at the end of the hurricane season. If they merged, the experts opined, the resulting megastorm would probably be a Category 7 or 8. If they didn't, and all made landfall, the destruction could even be worse.

Was the cluster of hurricanes somehow related to the Sun's aberrant behavior? Why was the Sun doing this to us? Not a scientific question, since the Sun has no mind or intent as far as we know, but a question that nonetheless was on everybody's mind. A little digging revealed that at least a dozen X-class solar storms had exploded during the bizarre and phenomenal week of September 7–13, 2005, making it the second most tumultuous period ever recorded on the Sun. No blackouts back then but plenty of hurricanes. Turns out that this immortally crazy week had come right after Katrina, and just before the even more powerful hurricanes Rita and Wilma, and also Stan, which killed more people than all the rest of them put together. For all the massive, relentless coverage of Katrina and her aftermath, why was it not revealed that what was one of the stormiest periods ever recorded on the face of the Earth had also been one of the stormiest periods ever recorded on the face of the Sun?

News teams were dispatched to the west coast of Senegal, where just offshore the hurricane cluster was stalled. Turns out this spot is where Katrina and most other Atlantic hurricanes are born. But where are they conceived? That's the question Carte Blanche, an enterprising CNN International team based in South Africa, decided to answer. Climatologists believe that hurricanes are spawned by atmospheric disturbances over the Sahel, an ecological corridor that divides the Sahara desert in the north from verdant land to the south. The Sahel is the closest thing to a "seam" that a continent can have, so the CNN International crew decided to follow that seam eastward across Africa, looking for telltale signs. A line drawn on the map of Africa from Dakar, Senegal, on the west coast all the way east follows the Sahel and terminates in the remote town of Boina, Ethiopia, which had been struck by an earthquake on September 14, 2005, the last day of the wild solar storms that had accompanied Katrina, Rita, Stan, and Wilma. Carte Blanche pointed their camera into the immense black fissure created by the Boina earthquake and reported with meticulous accuracy that the most esteemed geologists in the field now concur that said fissure signals the cracking apart of the African continent, probably right along the Sahel corridor seam, and the birth of a new ocean basin.

"Sun storms have been positively connected to hurricanes, earthquakes, and power outages. Such storms are widely expected, by a global consensus of astronomers, to climax at the end of 2012. How on Earth should we prepare?" Carte Blanche, and damn near every other news team on Earth, wanted to answer.

Government scientists stammered lengthy and unconvincing explanations about why the whole Katrina / sun storm / earthquake thing was probably all just a coincidence. As for the current leap in sun storms, well, there were lots of open questions. But journalists would not let the matter drop. The fact that 2007–2008 had been the International Heliophysical Year (IHY), in which thousands of scientists in dozens of countries had engaged in a multibillion-dollar research effort to learn more about the Sun and the threats it can pose, though entirely a matter of public record, seemed like an exposé when revealed during those tumultuous weeks of autumn 2012. Why had this collaboration been kept virtually secret? What were the astronomers trying to hide?

Then the cluster of hurricanes dissipated. Or maybe they just decided to come back later on.

Panic and confusion over the dark promise of the Mayan end-date of 12/21/12 proved rich soil for allegations of conspiracy, cover-up, and ineptitude. Establishment scientists who were long accustomed to kid-glove treatment by the media and politicians, and whose job when interviewed was simply to explain without having to defend, found themselves set upon as spies and traitors. Researchers from far-flung universities and obscure foreign institutes, independent scientists and inventors, science journalists and natural philosophers, united in this moment to launch a counterattack, a grudge match, against the science-as-usual bureaucrats. Rage against this doggedly unemotional group of professionals spilled like lava in every direction.

Scientists soon became the scapegoats du jour, blamed for just about everything in the world gone terribly wrong. Just as doctors, lawyers, journalists, and bankers had had their feet held to the fire by an angry populace frustrated with their arrogance, incompetence, and lack of accountability, now it was the scientists' turn to squirm. A multipart *New York Times* exposé spat lighter fluid into the flames, showing how hundreds of billions of taxpayers' dollars had been wasted over the past two decades by physicists trying and failing to control nuclear fusion, the most powerful force we know of in the Universe, the one that fuels the Sun and hydrogen bombs.

Scientists quickly replaced lawyers as the number one butt of jokes:

Q: "How many scientists does it take to change a lightbulb?"
A: "Depends on the size of the government grant."

But the real joke was that the scientific establishment, for the most part not guilty of the malfeasance with which it was charged, began lying their collective heads off, ostensibly for the public's own good, as time counted down to 12/21/12.

NOAH RETURNS

Sylvia C. Browne, a renowned psychic, seeks to dispel fears about the possibility of apocalypse happening in 2012 by beginning her recent book, *End of Days: Reflections and Prophecies About the End of the World* (2008), with a long list of errant doom prophecies. Popes, philosophers, scientists, and other assorted luminaries have erroneously advised us to hit the deck, including Christopher Columbus, who wrote in *The Book of Prophecies* that the world would end in 1658. Browne's list would have been a lot more reassuring had she not begun it with a quotation from an ancient Assyrian tablet: "Our earth is degenerate these days. There are signs that it is speedily coming to an end."

According to Browne's research, the tablet dates back to "approximately 2800 BC," which is plenty close to May 10, 2807 BC, the date that Bruce Masse, an environmental archaeologist at Los Alamos National Laboratory in New Mexico, now assigns to the beginning of the Great Flood, à la Noah and his ark. Masse is part of the Holocene Impact Working Group, an ad hoc alliance of scientists from the United States, Russia, Australia, and Europe devoted to studying the effect of extraterrestrial impacts on the Earth. ("Holocene" is a geological term meaning recent or current epoch.) The Holocene Group is literally rewriting the history of human civilization to include a lot more major impacts from outer space. It has spectacularly demonstrated that a comet impact caused the Great Flood some 4,800 years ago, affecting most of the world.

According to the Bible,

> Wild animals of every kind, cattle of every kind, reptiles of every kind that move upon the ground, and birds of every kind—all came to Noah in the ark, two by two of all creatures that had life in them. Those which came were one male and one female of all living things; they came in as God had commanded Noah, and the Lord closed the door on him. The flood continued upon the earth for forty days, and the waters swelled and lifted up the ark so that it rose high above the ground. They swelled and increased over the earth, and the ark floated on the surface of the waters. More and more the waters increased over the earth until they covered the high mountains everywhere under the heavens." (Genesis 7:14–20)

The Great Flood investigation started when Masse noticed that many mythologies from around the world described similar conditions—months of rain, great flooding, and devastation. Second only to the story of Noah, the best known of these myths is from Plato, who in his *Critias* dialogue (111.e.5–112.a.4) referred to the three catastrophic floods. The greatest came at the time of Deucalion, who with his wife, Pyrrha, built an ark after being warned by his father, Prometheus, that Zeus, the supreme Greek god, was about to punish humanity with an annihilating flood. Given the remarkable similarity of the stories, scholars have long considered God and Zeus, and also Noah and Deucalion, parallel figures. Also included in the group is Utnapishtim, the lone survivor of the Sumerian flood. His story was told in *The Epic of Gilgamesh*.

A few coincidences don't prove anything, so Masse set about gathering myths from around the world. A thousand or so of them, representing several hundred individual cultures, have been translated into English, and about half of them tell of a torrential downpour at a very early point in their history. A third of them, mostly from coastal cultures, also describe a tsunami occurring at that time.

"Because of astronomical and seasonal information encoded in the myths, as well as details about the deluge itself, it is virtually certain that the great majority of these myths represent a single event or simultaneous events," declared Masse, presenting his findings at the Tunguska 2008 conference held June 30 to July 2 in Krasnoyarsk, Siberia. The conference was organized to commemorate the 100th anniversary of the asteroid impact that flattened an area of about 2,000 square kilometers (775 square miles) in Tunguska, northern Siberia, making it one of the largest impacts in recent history. Considered a once-in-a-century event, the asteroid believed to have struck Tunguska was about the size of a ten-story building, much the same size as the one that just missed the Earth in March 2009.

Many of the ancient myths examined by Masse also mentioned a great horned beast appearing in the sky just before this event. Turns out that when a comet approaches from a certain angle to the Sun, its tail flips up and the net effect is that it looks just like a horned beast. Masse reported:

> Based on the descriptions of giant, elongated, and fiery or bright celestial supernatural beings in the deluge myths prior to or at the beginning of the flood storm, it is clear that the impactor was a comet and was visible to many or most cultural groups for several days prior to impact ... People in India observed the atmospheric debris plume and evocatively described it as doomsday clouds that look like a herd of elephants, emitting lightning, roaring loudly ... Several Indian myths describe fiery particles falling from the sky at the beginning of the deluge, as do myths from the Congo region of Africa and New Guinea, a seeming referent to the ballistic reentry though the atmosphere of superheated ejecta [that which is tossed out of a crater by the explosion that creates it].

AFTERMATH

Masse mapped it all out with his supercomputer and calculated that the comet must have hit somewhere in the Indian Ocean, so he circulated his data among his colleagues in the Holocene Group.

Dallas Abbott, a geoscientist at Columbia University's famed Lamont-Doherty Research Laboratory in Palisades, New York, had been using Google Earth and other mapping data to locate dozens of impact sites around the world. When the impacts occur on land, they create craters that when examined from above the Earth's surface look like bullet holes in glass. Eventually many of these craters fill up with water and become lakes. But when impacts are in the ocean or sea, the craters that result are at the sea bottom, making them much harder to spot. Telltale signs are therefore needed to guide the search.

All impacts from outer space create distinctive patterns known as chevrons. Here's how to make a chevron. Take a big stone, walk ankle deep into the waves at the beach, turn your back to the ocean, and throw the stone down into the shallowest part of the water with all your might. The sand that splashes up and out onto the shore is a chevron. By throwing down the stone in this manner, one creates a microtsunami, with the wave of water pushing the sand ahead of it toward the shore. If you repeated the experiment over and over again, the chevrons created would show certain structural similarities, such as forming a spray pattern fanning out from the point of impact. Different angles of impact would lead to different chevron results. If you threw the stone down hard enough, chemical analysis of the chevron would reveal little bits of the stone had gotten into it. If you turned around to face the ocean and threw down your stone into the water, a chevron would be formed, but would quickly dissolve without a trace because of the action of the waves.

Abbott sifted through her database and quickly came up with the smoking gun: four immense wedge-shaped chevrons at the southern shore of Madagascar, a strange and wonderful island 300 miles east of southern Africa. It's as though Noah built a special ark for Madagascar alone, since half of its birds, reptiles, and amphibians are unique to the island, as are all of its lemurs, the graceful, leaping cousin of monkeys, apes, and humans, believed by some, because the lemur stands erect, to be Darwin's "missing link" between primates and people. There being no crater on the island to account for the impact, Abbott and her team analyzed the size and configuration of the chevrons and extrapolated the approximate location, force, and trajectory of the impact that spat them up on shore. She and her colleague Lloyd Burckle then zeroed in on the site with scientific research satellites, and in 2005 they discovered, some 900 miles southeast of the Madagascar chevrons, the Burckle crater, 18 miles in diameter, 12,500 feet below the surface of the Indian Ocean.

The Holocene Group now believes that the comet that smashed into the Indian Ocean some 4,800 years ago produced a tsunami approximately 200 meters (650 feet) high, at least ten times the size of the killer tidal wave that swept across those same waters on December 26, 2004, inundating the coastlines of Indonesia, India, Thailand, Sri Lanka, and Myanmar, and penetrating inland up to half a mile (roughly one kilometer) or more. By contrast, the tsunami of 2807 BC flooded out all the coastlines to a depth of 2.5 miles (4 kilometers) or more. Had the 2004 tsunami been that large, deaths would have totaled not in the hundreds of thousands but in the tens of millions.

Because the Burckle comet came down in the middle of the ocean, people were probably spared from being burned alive by the impact fireball, which would have extended outward about 1,000 kilometers from the crash site. The shockwave must have blown down most trees in a 2,000-kilometer radius, though again, the midocean crash point served to spare all but Madagascar from severe damage. The spray of ejecta, however, is calculated to have extended some 9,000 kilometers, almost a quarter of the way around the world, including much of Africa, the Arabian Peninsula, India, mainland Southeast Asia, Indonesia, and Australia.

Heat from the monstrous impact of 2807 BC vaporized millions of tons of seawater that condensed over the course of several days.

"During this time that the water was cooling, global winds would transport the clouds thousands of kilometers. Thus, an impact of this size could be the source of deluge legends in the continents that circle the Indian Ocean: Africa, Australia, Europe and Asia," write the Holocene Group authors in "Burckle Abyssal Impact Crater: Did This Impact Produce a Global Deluge?" by D. H. Abbott, L. Burckle, and P. Gerard-Little of Columbia University, W. Bruce Masse of Los Alamos National Laboratory, and D. Breger of Drexel University.

Although the Indian Ocean impact could account for the massive rainfall that flooded out the region, deluge legends from that time period extend far beyond the Burckle circle. To fully account for the ancient flood reports, the Holocene Group now hypothesizes that the comet that caused the Great Flood must, upon entering the atmosphere, have broken up into at least three pieces: the one that hit the Indian Ocean, a second that struck in the eastern Pacific near the equator, and a third that hit in the far northwest Pacific. The Holocene Group is now searching for these craters. The Holocene Group reports:

> The archaeological record of this time period is consistent with a worldwide deluge catastrophe, including reductions in population, major movements of people, a proliferation of new languages and dialects, and settlement patterns stressing the use of higher elevation topography, including the construction of massive mounds. There is paleoenvironmental data indicating the sudden appearance of new savannahs and grasslands where forests once stood . . . The indirect effects of the Burckle crater event and associated processes would have been more devastating to humanity than the actual impact

itself... The deluge myths describe starvation and human suffering during and after the event.

Marie-Agnès Courty, a French soil scientist at the European Center for Prehistoric Research, has unearthed compelling evidence to corroborate the Great Flood theory. Her examination of fossilized soil samples from around the world confirmed the worldwide distribution of cosmogenic (originating in outer space) particles deposited by an impact some 4,800 years ago, precisely the timeline for the Great Flood. And Dee Breger, a microscopist at Drexel University in Philadelphia who works with the Holocene Group, analyzed samples of the Madagascar chevrons and found that, fused to the foraminifera, tiny ocean fossils contained therein, were iron, nickel, and chromium, the heavy metals typically deposited by extraterrestrial impacts. Moreover, those three metals were found in the same relative proportion as would be the case when a chondritic (lumpy) meteorite, the most common type, vaporizes upon impact in the ocean, according to *The New York Times* (November 14, 2006).

The Holocene Group's deeper message, I feel, is to respect the very real possibility that some ancient legends are not just fairy tales but lyrical renderings of truths that, scientifically analyzed, may hold vital, life-or-death information for us today. A good story but a bad omen, that the truth behind the Great Flood should finally be unearthed a few scant years before apocalypse is once again in the air. Of course, if the ancients survived, so certainly will we, with our infinitely superior knowledge, wealth, and technology. But with a thousand times more people and structures on the planet today, the suffering will be exorbitant.

Are there any signs, hints, or tip-offs that a cataclysmic impact might be imminent?

The spring of 2807 BC was a very lively time, cosmically speaking, according to the Holocene Group's careful reconstruction. Comparisons of myth details with astronomy software reconstructions indicate that "several unusual celestial phenomena, including extraordinary conjunctions of planets and eclipses of the Sun and the Moon, occurred at around the time of the impact. The effect of these observations, when coupled with the impact itself, profoundly influenced religious beliefs and also did much to foster a critical interest in celestial phenomena that is reflected in a flurry of activity evident in both early astronomy and astrology during the middle of the third millennium BC."

Rare planetary conjunctions, eclipses of the Moon and the Sun that perhaps fulfill Paul's prophecy in Acts of the Apostle, will, as noted in the Introduction, also characterize the autumn of 2012. Do the similarities end there, or will a catastrophic comet impact also occur in that perhaps fateful year?

RETHINKING OUR ASTEROID/ COMET DEFENSE

Astronomers have become quite good at tracking NEOs (near Earth objects), and the good news is that none (that we know of or have been told of) is scheduled to come close to colliding with the Earth in 2012 or any other time soon. The Spaceguard Foundation, established in Rome in 1996 and now comprising a network of observatories around the world, diligently searches the skies for such threats. Those working with the project are pretty confident that they have identified and calculated the orbits of all asteroids large enough to do significant damage to our civilization. Same pretty much holds for short-period comets, which originate in the Kuiper Belt that extends from Neptune past Pluto. Short-period comets have orbits that last less than two hundred years and are therefore pretty easy to track. Halley's comet, the Old Faithful of the comet system, is probably the best known of the short-period comets, returning every seventy-six years.

There is a chink in our space defense surveillance system, however. It comes in the form of long-period comets that originate in the Oort-Opik Cloud at the very edge of the Solar System. Long-period orbits take more than two hundred years, sometimes many more. It is therefore entirely possible that a long-period comet could all of a sudden descend murderously out of the void.

In a literal, physical sense, it would be the Earth's own fault. It is not really accurate to say that NEOs bombard our planet. Rather, like any other planet, the Earth sucks in the NEOs with its immensely more powerful gravitational field. About 100 tons of space debris per day are pulled into the Earth's atmosphere, which, like a membrane, refines most of these space chunks into dust and/or vapor that eventually makes its way to the ground.

Over the eons, the Earth has benefited mightily from all this space-sucking. A number of astronomers now believe that the ice crystals in comets' nuclei have been a significant source of water for our planet. Current guesstimates are that comets provide as much as an inch of water every ten thousand years, which would work out to a layer 7.5 miles deep since the Earth was born. Fortunately for us landlubbers, most of this water was either out-gassed before the atmosphere formed sufficiently, bound up in organisms, or dissociated by chemical processes such as photosynthesis and weathering.

Asteroids and meteorites have also supplied valuable materials, perhaps including a significant share of the iron ore on the surface of the Earth, which saw most of the iron it was originally endowed with migrate to the molten core during the fiery first billion years after creation. That's probably where most of our iridium went as well. Chemically similar to platinum and ten times as rare, iridium gravitates toward iron; large amounts of it are therefore believed to have migrated core-ward during the early days of the planet's formation. Were it not for as-teroid and meteorite impacts, this extraordinarily dense, stiff metal with extremely high melting and boiling points, would be present on Earth's surface only in trace amounts. Without these periodic injections of irid-ium from outer space, we would have had a devil of a time manufacturing everything from spark plugs to semiconductors to electrical generators for unmanned spacecraft, along with many other advanced devices that require the most corrosion-resistant metal known to man.

The problems come when the Earth bites off more than it can chew, sucking in an object that its atmosphere cannot completely digest, such as the iridium-rich asteroid that extinguished the dinosaurs 62 million years ago. But even the worst cataclysms have their bright side. Ever since Stephen Jay Gould advanced his theory of punctuated equilibrium, amending Darwin's stately view of evolutionary progress to one that is periodically shaken up by extraterrestrial impacts and other cataclysms, scientists have come to understand that, if anything, these cosmic smacks to the head hasten evolution, causing the global ecosystem to recover rel-atively quickly (at least compared to the 5 billion years the Earth has been around) and flourish at a higher level than ever before—much the way that pruning a healthy plant will soon cause it to bloom more gloriously than ever.

Overall, the Spaceguard Foundation pegs the frequency of globally catastrophic impacts at between every 100,000 to 10 million years. Since the Burckle/Great Flood impact happened less than 5,000 years ago, does that mean we're safe for at least another 95,000-plus years? Or as *The New York Times* observed in its article on the Holocene Group, do catastrophic impacts actually occur far more frequently than space scientists have tra-ditionally believed? Indeed, for every chevron and impact crater discov-ered on land, which covers about 29 percent of the Earth's surface, the

Holocene Group must increase their overall impact estimates enough to factor in the probability that many more such sites lie somewhere beneath the 71 percent of the surface covered by the sea. The group of scientists now believes that impacts with the potential to inflict broadscale catastrophe occur perhaps as often as every 1,000 years.

Our whole way of thinking about impacts and their consequences has to be retooled. Where an impact occurs on the Earth's surface is more important than ever, because the population increases with each passing year, as does the level of socioeconomic interdependence among regions. What if the equivalent of one of the three chunks of the Burckle comet landed today not in the Indian Ocean but in the Mediterranean Sea? Though the seismic and meteorological consequences probably would not qualify as a global disaster on the scale of the Great Flood, the carnage caused in the heavily populated, highly developed Mediterranean region, the damage to Italy, France, and Spain, some of the world's most cherished and productive societies, and the potential for conflict in the eastern Mediterranean, where Israel, Egypt, Syria, the Palestinian Authority, and Lebanon share shores, could grievously debilitate civilization.

The further civilization advances, the more susceptible it becomes to primitive impacts from the sky. Such thinking transforms the calculus of catastrophe prevention, forcing us to consider seriously the trillion-dollar investment that would likely be required to build a defense system that would track all NEOs and neutralize them when they threaten the Earth. Sadly, there is no guarantee that such a weapon would work properly, given the frankly dismal results we have seen in the arena of antimissile technology, where one missile is fired to knock out another missile that has already been launched. There have been a lot more misses than hits, even though human beings control both target and interceptor. Worse, our asteroid/comet weapon could work imperfectly, not obliterating a NEO but breaking it up into lethal chunks that smash variously around the world, which apparently happened to the comet that caused the Great Flood. And who's to say that the antiasteroid technology wouldn't be hijacked, for dastardly this-world attacks?

As for schemes to divert incoming comets and asteroids so that they miss our planet, the Spaceguard Foundation points out that the amount of force required to accomplish this varies inversely with the square of the period of time that will elapse before impact. Thus, when objects are far away, at least fifty to a hundred years out, they can be effectively diverted by slight nudges from spacecraft well within our capabilities to send. Redirecting the object even a fraction of a degree when it's that far off creates an error that multiplies the farther it travels. But once incoming objects bear down at close range, their trajectory must be profoundly altered, requiring the intercepting spacecraft to deliver a truly astronomical punch, though of course without breaking up the asteroid or comet such that it showers us, Burckle comet–style, with lethal chunks too big for the atmosphere to digest.

Bottom line, the need for a NEO-defense system has never been greater, but the chances of our actually building one are slim to none. In today's economic emergency, taxpayers will not fork over the vast sums required, particularly for an interception system geared to objects fifty or a hundred years away. A blow-'em-out-of-the-sky shield might have more popular appeal. However, enormous political pressure would be required to sell the idea to Russia and China, who would likely consider such a system in violation of antiballistic missile treaties and potentially directed against their own satellites and/or missiles.

The whole NEO-defense issue would create a god-awful mess that could take a generation to resolve. Not that it would be worth the trillion dollars, but moving forward publicly with such a weapons system would raise some fascinating issues about attitudes toward such catastrophic impacts, including whether or not they are punishments we deserve.

BECAUSE GOD WAS ANGRY

All but forgotten in the Holocene Group's data-sifting through deluge myths were the explanations as to why the Great Flood occurred: because God was angry. This holds true in both the story of Genesis and in Greek mythology, where Zeus sent the flood to punish the people for their irreverence. God's anger also caused the flood in the Sumerian tales recounted in *Gilgamesh*, and no doubt in many other mythologies.

Logically, there is no case at all for arguing that the deity, should any such entity exist, has anything to do with the extraterrestrial impacts that have periodically stunned the Earth over the past 5 billion years. If God hurls comets and asteroids down upon the Earth to punish humanity for its evil ways, then why has He/She been pummeling the Earth for billions of years before human beings came on the scene? Target practice? Because the dinosaurs were behaving badly?

One could speculate that certain of these catastrophes, including the Great Flood, were of divine origin because that's what's written in the Bible and other sacred ancient texts. However, there is no corroborating evidence to support a connection between the moral quality of human behavior and the occurrence of natural disasters. Only the most primitive superstition might prompt one to wonder if the level of conduct in our current civilization is abysmal enough to tempt God to toss a molten one at our head. Likewise, the straw-grasp that if we were somehow to improve our behavior and pray our heads off, perhaps He/She might be persuaded to whiz the beanball by our ear instead.

We modern, rational, scientific thinkers graciously excuse the primitive ancients for attributing human characteristics to natural phenomena, and for thereby concluding that the Great Flood was a product of God's rage. We dismiss their explanation as unscientific, which of course it is, though it does touch a nerve. What are we, as a civilization, guilty of? Viewed as a collection of individuals, each of whom has certain inalienable rights, civilization would never deserve punishment because far too many good people and innocent children would suffer and die in the process. But viewed as a coherent, organismic entity, of which all individuals are but a part, civilization is no more or less deserving of chastisement than you or I.

So how much punishment do we deserve? How much do we need?

Aftermath Scenario: Comet Strikes the Mediterranean

Had the comet smashed into, say, the Caribbean, or the North Sea, the anguish and excitement would have been much easier to contain. The postmortems would have been primarily scientific in tenor, and the rescue and relief operations would not have been nearly so complicated and politically charged. However, the fact that it hit the Mediterranean, birthplace of Christianity and Judaism, central to Islam, just made it seem as though God was raining down His wrath upon humanity.

When what later came to be known as Noah's comet first appeared on November 13, 2012, the day of a total solar eclipse, only the sky buffs seemed to care. Several nights later, the faint white smudge was glowing like a lone Christmas bulb on an otherwise bare tree, a beautiful cynosure, though not exactly festive. The general public was still jittery from all the power outages and communications snafus caused by the pummeling of solar storms, and was in no mood for any more surprises from the sky. Reports of odd occurrences flooded the Internet and the popular media. Boats rocking in harbors where there were no discernible waves. Animals who never left their all-fours walking on hind legs. Seers having comet visions that caused their bodies to writhe like snakes. Most, if not all, of these stories were trumped up or overblown. But they got under people's skin anyway.

Government officials from many nations, acting with surprising speed and sensitivity, put together a road show of unimpeachably credentialed astronauts, Nobel luminaries, and astronomers whose job it was to explain that the comet was not going to hit the Earth and calm people's fears. (Religious leaders were briefly considered for inclusion in the panel, but it was decided that that would seem too apocalyptic.) What might otherwise have gone down as a case study in the annals of good government ran into a snag: the scientists couldn't rule out the possibility of an impact. No problem, replied the comet campaign organizers, to whom sincerity was but a style; just fudge the statistics a little and draw the charts and graphs with a happy face in mind.

The scientists were won over by the rational, humane, plausible argument that the panic caused by truthfully relating that Noah's comet might be heading toward Earth would cause far greater harm than would the lie that the world was safe. Civil and military rescue operations would of course go on the highest alert, just as a precaution to calm people's nerves, it would be explained to the media. There wouldn't be much in the way of actual strategic deployment anyway, because no one could predict with any degree of certainty where on the globe the impact(s) would occur.

Heroically, the esteemed panel of scientists put their duty as citizens of the world, as human beings, above their lifelong dedication to factual truth, held a global press conference at the United Nations watched by almost as many people as watched the last year's Super Bowl, and immediately set off bullshit detectors around the globe. Scientists make lousy liars. Who can blame them? Unlike politicians, whose predilection to deceive occasionally dovetails with their legitimate commitment to protect the public from information that might lead to a panic, and unlike artists and entertainers, whose only job, really, is to mesmerize, the duty of a scientist is to lay out the cold hard facts (or, in this case, the molten ones) of the physical world. Not that the Committee of Scientists was ever caught in a lie or factual contradiction, just that their facial expressions and tones of voice reeked of guilt and fear.

Shortly after the end of the disastrous press conference, a breakaway group of scientists hastily reassembled the media and urged people around the world to evacuate the coasts. Two or three kilometers (a mile to a mile and a half) should do it, if any of the tsunamis created were the size of the one that swept the Indian Ocean in December 2004. Expert opinion was divided on whether the worst-case scenario was for the comet to smash the surface intact or for it to break up into half a dozen or so chunks, all of which would hit different spots around the globe. However it came down, the impact points considered least disastrous to global civilization were remote and sparsely populated stretches of land, such as Siberia and the Sahara desert. The greatest catastrophes would occur if the comet hit deep sea—the deeper the water, the bigger the splash, meaning the larger the tsunami and the more water that would be vaporized, soon to come back down in torrential, Great Flood–type rains.

To the scientists, the debate about the details of the comet impact was of secondary importance, since nothing could be done about it anyway. That's where the religious leaders stepped in.

"Our Father who art in Heaven, hallowed be Thy name. Thy kingdom come, Thy will be done on Earth as it is in Heaven."

Although Jesus taught us the Lord's Prayer, it makes no mention of the Trinity, or of any other doctrine objectionable to the other great faiths of the world. Intriguingly, the person intoning this prayer starts by beseeching God to do God's will. There's a certain circularity to this reasoning. It's like praying for what's going to happen anyway. The deity, like any other entity, will certainly endeavor to do what it wants to do. Placing God's will above one's own, even if there is a conflict, is nonetheless a healthy exercise in humility. Perhaps there is scope to God's will, and, just possibly, fervent prayer can influence outcomes within that scope. Billions upon billions of prayers were sent heavenward in that hope.

As Noah's comet filled the sky, it was as though the German poet Rainer Maria Rilke rose from his grave: "Then beauty is nothing but the beginning of terror" became the supreme truth of that moment of existence. Humanity split into ants and grasshoppers: those who buried themselves and their young in the best shelter they could find, and those who partied their heads off. Everyone without an abnormally large death wish hoped and prayed that Noah would burn up completely, or explode into sparks that would fall harmlessly to the ground. That's pretty much what happened on the day of impact, November 28, 2012, except for one small hardcore chunk of the nucleus, which remained intact as it plunged into Cape Matapan, in southwestern Greece, three miles north of the cave known since ancient times as the home of Hades, god of the dead.

The blast devastated Sparta just to the north, and the shockwave blew down hundreds of structures around nearby parts of the eastern Mediterranean coast. The tsunami caused by the tremendous impact battered the islands of Crete, Cyprus, Malta, and Rhodes. Damages ran to hundreds of millions of dollars. Deaths, all told, numbered almost a thousand, appeasing Hades, some said. Tragic indeed, but compared to what could have been, thank you, God!

Had Noah's comet hit a few miles to the south, east, or west, it would have missed Cape Matapan and landed in the water, making the cataclysm unfathomably more horrible. The water around the Cape is the deepest part of the Mediterranean, almost 15,000 feet, deeper even than the spot where the Burckle comet landed in the Indian Ocean in 2807 BC. Tsunamis would have swept over the Mediterranean's 100-plus islands, and coastal cities from Barcelona to Beirut would not have fared much better. Casualty and damage figures would have increased tenfold, maybe a hundredfold.

And the Mediterranean, the most storied, cultured, glamorous, iconic million square miles on Earth, from the French Riviera to the shores of the Holy Land, would have been decimated.

Did the billions of prayers alter the comet's trajectory just enough to spare us from what would truly have been a megacatastrophe of a scale not seen since the Great Flood? Or was it all just a random occurrence? The deeper everyone got into the philosophical debate, the more confused they became. Except, that is, for religious scholars who noted that November 13, 2012, the date when the comet first appeared in the sky, corresponded with the Old Testament prophet Joel's prediction, as cited by Peter in Acts of the Apostles, just as the impact date of November 28 corresponded with the prophecy of the eclipse of the Blood Moon. But whether that meant that the biblically foretold end was nigh, or that we had been spared by God this time, was a matter of much debate, except, that is, to powerful religious fundamentalists, Muslims, Christians, and Jews united only in their obsession with Armageddon, the final war of Good versus Evil that would result in the end of humanity. To the Armageddonists, the comet crashing down in the Mediterranean on November 28, 2012, was a starter's pistol shot from Heaven above.

The Earth's atmosphere is a cocoon that has been incubating a very precious life form—some think of it as a superorganism called Gaia—for about 4 billion years. Every so often, the cocoon has been pierced by outer space impacts, exploded by volcanoes, and seared by solar blasts, but the damage was always repaired. Yet since the Industrial Revolution a century and a half ago, the cocoon has for the first time been under pressure from within. Intensifying bursts of heat and motion coinciding with the human population explosion have come in the form of global electrification, nuclear events, climatic warming, and natural catastrophes. Ozone holes are poking through the stratosphere. Pole-to-equator cracks are gaping through the Earth's protective magnetic field. Spacecraft regularly breach the planet's domain. At the outer fringes of the atmosphere, the human species has created an extensive satellite network, much like a biological network of neurons, that we have come to rely on for most of our distance communication. Our cocoon is beginning to fray. The superorganism inside is stirring, expanding, planning its escape.

An observer looking down upon the Earth might well find this whole process life-affirming, like a butterfly wriggling out of its silky blanket, or a hatchling bird pecking through its eggshell—a marvelous transformation to the stage in which the organism's life force is most gloriously expressed. However, the outlook is somewhat less positive from inside the cocoon. It's nice in here, although it is getting a little crowded, a little hot. Plus, who knows if this birthing process will succeed? A certain percentage of butterflies and chicks are stillborn, after all. If there are no guarantees of survival for conventional organisms, why should there be any guarantees for our superorganism, with the commensurately super-challenges we face as we burst into the unknown?

Interesting that the Mayan prophecy for 2012 is that it begins the Fifth Age, characterized not, as the previous four ages were, by any of the terrestrial elements of earth, air, fire, and water, but rather by ether, invisible and intangible, the stuff of which so much of the cosmos is made. Grade school science taught us that most of outer space is a vacuum, which is to say, nothing, no atoms. Somehow, though, light propagates effortlessly through this nothing at a constant 670,000,000 miles per hour, meaning therefore that time exists in the vacuum as well. As does probability, the cornerstone concept of quantum physics; why else would Albert Einstein, upon considering quantum theory, have irritably exclaimed that "God does not play dice with the world!" So if ether is taken to mean the basic substance of the cosmos, even setting aside the gazillions of stars, pulsars, planets, asteroids, comets, black holes, gamma bursts, and electromagnetic whirligigs, ether may be said to be devoid of matter—even energy, perhaps—but not of significance.

The essence of the great Mayan vision is ecstatic, with human civilization leaping in 2012 to a new and higher level of collective consciousness and mutual understanding. Recall that the Fifth Age, also called the Ethereal Age, is slated, just as the Mayan ages preceding it have been, to last 5,200 years, roughly until AD 7212. Beyond that, who cares? In fact, how many of us truly care about what happens beyond, say, AD 2100 or so? I've got two children under the age of ten who have a decent shot of living on into the twenty-second century, so I do care, in the abstract. But as for anything happening beyond that, it had just better take care of itself.

The working assumption is that the timing all works out, that the new planetary life form will not escape its cocoon before it is ready to handle its new environment. It will know instinctively, perhaps along with some guidance from a parent, how to move independently, find food, and avoid predators. If it does not survive, well, it must not have been fit enough to begin with, and its death, however saddening, is in the natural order of things.

Just as we are beginning to get comfortable with thinking globally, with grasping the ways in which trade, military security, telecommunications, and ecology all interrelate organismically on a planet-wide scale, the challenge is thrust upon us to think beyond globalism, and to understand that the next stage of our evolution is to explore, cultivate, and populate the Solar System. But we won't "boldly go" just because it is in the human nature to soar. We'll go for the same reason that most other immigrants do: because things aren't working out back home.

In the first section of this book, we examined events that might well occur in 2012, a space weather assault on the electrical power grid, military or terrorist attacks on our technological infrastructure, and a prophecy-fulfilling comet impact. In this section, we examine that year as a potential turning point for ongoing environmental and health threats. You will encounter two somewhat contradictory chapters, "Sweltering in Siberia," which takes seriously the climate change threat, and "Changing Climate Change," which takes seriously the threat that the environmentalist agenda to combat climate change could do more harm than good. Neither set of outcomes can be ruled out with confidence, so I have included both of them, not just for "equal time" purposes, but also in hopes of identifying solutions that encompass these divergent opinions.

SWELTERING IN SIBERIA

et's be honest, global warming has its pleasures. No jacket in the middle of winter, beach days after the leaves are off the trees. Not a big Sarah Palin fan, but come on, if you lived in Alaska, how upset would you be with milder temperatures, stranded polar bears notwithstanding? Not to say that the weather hasn't been a bit bizarre recently. Take winter 2008, which started out to be one of the coldest on record in Alaska until warm Chinook winds swept up from the south at 110 miles per hour, melting all the ice and snow and causing massive flooding in downtown Anchorage, right in the middle of January.

I've never been to Alaska, although I did visit Siberia in July 2008, and it turned out to be one of the hottest trips of my life, comparable to the June I once spent in the Amazon rain forest. In Siberia the summer sun burns from dawn to almost midnight, keeping temperatures at 90 to 100 degrees F (32 to 36 degrees C) for twelve or fourteen very humid hours per day. The buildings, designed to be cozy in the winter, are stifling in the summer. Air-conditioning is not a common amenity in Siberia, so there was no relief to be gotten from ducking into a bar or a restaurant, or even while incarcerated in my hotel room. But I was the only one complaining. To the last man, woman, and child, Siberians are gaga for whatever the opposite of winter is. People's attitudes toward global warming depend a lot on where they live. Had I taken a poll right then and there during those sweltering afternoons, global warming would have had a higher approval rating than chocolate. Warmer, longer summers and above-zero winters, whatever the cause, is something folks in cold-weather climates can definitely appreciate.

Deep down one might sense that all this climate craziness is man-made and therefore "unnatural," but that's what philosophers are for, to point out that human beings are part of nature and that whatever we do is, therefore, natural. Not a slam-dunk argument, exactly, but cogent enough to cloud the issue, since human beings are, in reality, both a part of nature and also the sole source of all activity deemed unnatural. That the Earth has apparently warmed and cooled periodically over the eons also lends some comfort; there are cycles to everything, turn, turn, turn. Climate change, we need to remember, is a natural, cyclical process that began when the Earth was created 5 billion years ago and will continue until the planet is no more. Just as species evolve, as the Solar System moves through the Milky Way, as the Sun burns through its various cycles and phases, so too changes the temperature, precipitation, and precise gaseous mix of our atmosphere. It could not be any other way. Mostly, this eternal, inevitable, life-sustaining process happens gradually, with nature and civilization adapting along the way. It would be the height of folly to try to stop it.

After some chardonnay on the patio on a balmy December evening, optimism comes easy: We, meaning human civilization, made it through the Little Ice Age, a 500-year cooling period that ended about two centuries ago, and we'll make it through this Little Warm-Up as well. Unless, of course, the warm-up takes 15 years, or even 50, to happen, instead of 500. That's what is called "abrupt climate change," defined in the United States Climate Change Science Program's report, issued in December

2008, as "a large-scale change in the climate system that takes place over a few decades or less, persists (or is anticipated to persist) for at least a few decades, and causes substantial disruptions in human and natural systems," write P. U. Clark and A. J. Weaver, authors of the summary.

Abrupt climate change (ACC) is what we rightly fear, because it could well be catastrophic. Every now and then the climate changes course, because of impacts from outer space, internal terrestrial processes we only dimly understand, and now, it seems, human activity. Unlike gradual climate change, which can be tracked over the centuries and millennia and therefore projected with some degree of confidence, ACC is, by definition, a break with the past. After the last Ice Age ended some eleven thousand years ago, global temperatures leapt 10 to 15 degrees C (20 to 30 degrees F) in a matter of decades, or so scientists currently believe. The fact of the matter is that we really have no way of knowing how many ACC episodes have happened before, since the ice core and fossil records on which so much climatology is based are not sufficiently calibrated to reliably indicate disruptions much shorter than a century. For example, Al Gore, for whom I have the utmost respect, used to give a presentation on global warming in which he presented results from a Vostok, Antarctica, ice core sample to the effect that a rise in carbon dioxide had caused a temperature increase. Later examination of the ice sample indicated, however, that the temperature increase actually preceded the uptick in carbon dioxide levels by half a millennium.

Even when accurately calculated, long-term climate change data can obscure ACC entirely. Which century would you rather live in: one where average global temperatures gradually rose from 50 degrees F to 50.5 to 51, or one where temperatures started at 50 degrees F, dropped to 47, jumped to 53, then ended up at 51? Both centuries would ultimately have warmed just one degree, but life under the second set of circumstances would have been exponentially more chaotic and dangerous.

Same problem holds for comparing a given century in which the hypothetical one-degree temperature rise was pretty much evenly distributed around the globe, with another century in which some regions leaped, some plummeted, still all averaging out to a one-degree global increase. Imagine, for a moment, that the Middle East, always the flash point, were to experience a sudden leap of several degrees C over the coming decade, parching agriculture, spiking energy demand, and pushing overcrowded cities and sectarian tensions that much closer to going molten. At the same time, temperatures might ease commensurately over the comparatively stable and less populous South Pacific region, leaving us with the hindsight conclusion that, climatologically, basically nothing happened.

Forecasting the climate via computer models is even harder than piecing together its history. The global ecosystem is laced with confusing feedback mechanisms, negative and positive. An example of a negative feedback system is when ocean temperatures rise and more water evaporates, making clouds, which then help lower ocean temperatures by preventing sunlight from reaching the surface. In general, negative feedback systems compensate for some, but not all, of the input that gave rise to them. So the cloud formation that occurs over the warming ocean would offset a portion of the rise in ocean temperatures. Seems simple enough, except for the fact that sometimes clouds actually do the opposite and warm the planet's surface. Anyone who has spent time in a wintry climate knows that the coldest winter nights are the clearest ones, without any blanket of clouds overhead. Thus, some climate models consider cloud formation as a cooling factor, others deem it a warming factor, and still others try to incorporate both possibilities.

Where do the clouds come from in the first place? Complicating the climate modeler's task even further is the uncertainty of the science behind meteorological events. For example, one school of thought holds that cosmic rays (essentially all emanations from outer space that do not come from the Sun) are instrumental in cloud formation; the more cosmic rays, the more clouds are formed. The stream of cosmic rays, however, is believed to vary inversely with the intensity of the solar wind, which is the stream of charged particles from the Sun. The heavier the solar wind, the less capable cosmic rays are of penetrating that wind. So in order for a climate modeler to predict accurately how cloudy a given period will be, he or she would first have to estimate the intensity of the solar wind that might dampen the stream of cosmic rays and thus reduce cloud formation.

Climate modelers usually work in time frames of thirty years or more, meaning that they need not target their predictions so precisely as to correspond, say, with the ups and downs of the Sun's eleven-year sunspot cycles, which correlate pretty closely with the intensity of the (cosmic ray–blocking) solar wind. However, solar activity does not slavishly follow any norm. There were virtually no sunspots, and a correspondingly faint solar wind, for the seventy-year period from 1645 to 1715. This period, known as the Maunder Minimum, came at the depths of what is known as the Little Ice Age, a cooling that settled into the Northern Hemisphere for a total of three hundred years or more. Climate modelers would never have been able to foresee either the drop-off in solar activity or its impact on Earth. Neither, for that matter, would the modelers have been able to foresee our overheated climate today, since, as previously noted, five of the most intense solar cycles on record have occurred in the last fifty years.

Bottom line, predicting when and how the climate will change abruptly is, well, damn near impossible. It's a lot to ask of any brain, computerized or otherwise, to forecast when a stable system will become unstable, when the rational will become irrational. Predicting abrupt climate change is kind of like predicting when a heart attack will actually occur. In point of fact, chaos theorists have made strides toward understanding the role of random instability in cardiac health, with the surprising finding that a lack of chaotic interludes may indicate the onset of certain types of myocardial infarctions. But foretelling when the global ecosystem might suffer an analogous geophysiological breakdown is infinitely more complex and challenging. Consider the failure of chaos theory to anticipate the breakdown of a much smaller mechanism, the global financial system. Top economists are quite familiar with the basic principles of randomness and discontinuity, and no doubt include the finest theoretical chaos software in their forecasting arsenal. But they were as stunned as the rest of us when Wall Street suddenly clutched at its chest in September 2008.

The scientific community is under enormous political pressure to come with answers concerning the timing and severity of ACC that it cannot reliably provide. What can be confidently extrapolated, however, are the ACC scenarios likeliest to occur. According to the 2008 U.S. Science Program report, these fall into four basic categories: (1) the melting of the polar ice caps, (2) the runaway desertification and depletion of fresh water supply, (3) the diversion/depletion of the Gulf Stream, which warms eastern North America and Western Europe, and (4) the sudden, rapid burst of certain greenhouse gases into the atmosphere. Of these, the 2008 report found that the most pressing current threat of ACC comes from the melting of the polar ice caps.

GLOBAL COOLING

"The North Pole is melting . . . The last time scientists can be certain that the pole was awash in water was more than 50 million years ago." Ever since The New York Times (August 19, 2000) ran John Noble Wilford's much maligned story on its front page, climate change naysayers have been merciless. The source of Wilford's information was none other than James J. McCarthy, a Harvard oceanographer who was on a summer tourist cruise to the pole, became astonished at seeing all the open water at the pole, and, frankly, got carried away in the conclusions he drew. At the time, McCarthy was cochair of the United Nations Intergovernmental Panel on Climate Change (IPCC) section on "adaptation and impacts" of climate change. What a gift to the climate change naysayers! Harvard, the UN, and the *Times*, the triumvirate of liberal wrongheadedness, caught perpetrating an alarmist, refutable, junk-science fraud.

Patrick J. Michaels, an environmental scientist with the Cato Institute, known for its indefatigable opposition to anything, especially environmental regulations, that would inhibit free market caprice, sliced into McCarthy's allegations. Citing evidence indicating that temperatures at the North Pole were no higher than they had been for several decades in the early twentieth century, Michaels asserted that there had been no net change in summer North Pole temperatures since 1930 or so. What's more, "most climatologists think the period from 4,000 to 7,000 years ago averaged at least 2 C warmer than the current era at high latitude. That's three millennia," Michaels wrote.

As a result of attacks by Michaels and others, the Grey Lady blanched and published a retraction of sorts, allowing that open water in the dead of summer is not all that rare, even at the North Pole.

But the underlying message of the *Times* article was spot-on. According to the 2008 United States Climate Change Science Program's report, conceived, prepared, and published entirely under the auspices of the Bush-Cheney administration, known for routinely minimizing the importance of global warming, the melting of the polar ice caps is leading to an unexpectedly rapid rise in sea levels, faster even than the 7 to 23 inches (18 to 58 centimeters) by 2100 previously predicted in the IPCC

AR4 study, *Climate Change 2007: The Physical Science Basis*, until then considered the most alarmist scientific study on the subject. (Citing the inadequacy of computer models and other scholarly deficiencies, the 2008 report declines to make specific projections on exactly how much sea levels will rise in the foreseeable future, just that they will rise more than predicted by the IPCC, which shared the 2006 Nobel Peace Prize with Al Gore for its work on climate change.)

One might think it comparatively straightforward to calculate the rise in sea levels that will be caused by glacial melting: simply estimating the amount of ice, then adjusting for different levels of ice density (the ice at the bottom of an iceberg is compressed by the weight of the ice on top of it). That result can be converted to free water volume, which is then added to the amount of water already extant, with sea-level estimates upwardly adjusted to account for the inrush of ice melt. But the fact of the matter is that estimates of how much the seas will swell from melting are all over the board, literally from millimeters to meters per century, that is, from barely noticeable increases readily defended against by properly securing the coasts with levees and sea walls, as should have been done in New Orleans, to New Orleans–scale disasters occurring every few years, regardless of whatever civil engineering efforts are undertaken.

What complicates sea level calculations is the very simple fact that ice displaces its own volume when it floats in water. Water, you might recall from middle school chemistry, is the only common substance that expands when it freezes. Everything else contracts when frozen. Try the following experiment: Take a glass, pour some water in it, then add some ice cubes. Mark with a crayon the level of the water in the glass, then set the glass aside to let the ice cubes melt. Examine the new water level. It should be at, or quite close to, the original crayon mark. Oceanographers, of course, are well aware of this elementary fact, though they vary widely in how they factor it into their estimates.

Clearly, sea levels are rising. Just ask anyone who lives in Indonesia, the Maldives, or any of the other island nations that find themselves increasingly swamped. Glacial melting is just as clearly a contributor to this situation. Moreover, the melting is accelerating, with net ice losses in Greenland leaping from zero in the mid-1990s to 100 gigatons per year in 2000 and 200 gigatons per year in 2006, according to the United States Climate Change Science Program.

"The best estimate of the current (2007) mass balance of small glaciers and ice caps is a loss that is at least three times greater (380 to 400 Gt a-1) [gigatons per year] than the net loss that has been characteristic since the mid-19th century . . . Significant changes in ice-shelf thickness are most readily caused in basal melting induced by oceanic warming. The interaction of the warm waters with the periphery of the large ice sheets represents one of the most significant possibilities for future rapid changes in the climate system," says the 2008 study.

The situation in Antarctica is even worse, according to Colin Summerhayes, executive director of Britain's Scientific Committee on Antarctic Research, which collaborates closely with the massive International Polar Year (2007–2008) research effort on changes at the Earth's poles. The largest western Antarctic glaciers are shrinking faster than had ever been expected, the loss being roughly equivalent to the meltdown of the whole Greenland ice sheet. This would add another 4 to 8 inches (10 to 20 centimeters) to the IPCC sea-level projections, which, as noted, already underestimate the run-off from Greenland. Worse, the entire western Antarctic ice sheet may be on the verge of collapse, a catastrophe that some experts claim would pump up sea levels an additional 4 to 5 feet (1.0 to 1.5 meters)!

Imagine swimming in the ocean and being caught between two waves, each coming at you from the opposite direction—in this case, one from the North Pole and the other from the South Pole. The threat of rising sea levels caused by polar meltdown inspires tremendous fear because the consequences are simple to visualize: island nations submerged and great coastal cities flooded out. Any number of nightmare scenarios have been generated whereby low-lying areas such as New York City, London, and Florida are devastated by the effluent of polar melt.

Even less understood, and just as dangerous in the long run, is the abrupt global cooling that will result from a meltdown of the polar ice cap. Global cooling might initially seem less than threatening, even a welcome antidote to global warming. From the perspective of the Gaia hypothesis that the Earth is akin to a living organism, the polar meltdown is simply an example of the global ecosystem regulating itself, cooling in response to having been overheated, kind of like the way one's body sweats to cool off. Fine for Gaia, with her million-year timescales, but not for us.

The worst storms are created by abrupt climate change, such as the cooling caused by ice melt. According to Ian Allison, also of the Scientific Committee on Antarctic Research, extremely large storms that now happen once a year will begin to happen on a weekly basis as sea levels rise to the levels now being predicted. Ted Bryant, a geomorphologist at the University of Wollongong in New South Wales, Australia, concurs, explaining that although warming is the underlying problem, the worst natural catastrophes actually tend to occur when the climate cools. "Climate disasters tend to happen when the climate is more variable, as it is today. In fact, the greatest disasters come as the climate cools because the cooling process essentially entails a release of heat and energy, all too often in the form of catastrophes," observes Bryant.

That glacial meltdown is a mortal threat to civilization has long been preached by Ervin Laszlo, seventy-six. "The frigid waters from the melting of Greenland will blunt the Gulf Stream, and Europe will be plunged into a freeze. The melting of the Greenland ice cap is the single greatest threat to the habitability of Western Europe, with extreme political, social and economic repercussions for the rest of the world. Cover 2% of the Sahara with solar panels. Europe's energy needs will be met and the crisis may be averted," declared Laszlo, author of *The Chaos Point*, in his March 2008 address to the capacity crowd at The Hague, hosted by Princess Irene of the Netherlands.

A systems theorist best known for his "connectivity hypothesis" integrating science and mysticism, Laszlo is founder of the Club of Budapest, whose members include Sir Arthur C. Clarke, the Dalai Lama, Jane Goodall, Mtislav Rostropovich, Muhammad Yunus (Nobel peace laureate and microcapital investment exponent), Václav Havel, Vigdis Finnbog-gadottir (four-time president of Iceland), Maurice Béjart (iconoclastic choreographer), Mikhail Gorbachev, Mihaly Csikszentmihalyi (flow theorist), Peter Gabriel, and Oscar Arias (Nobel peace laureate and current president of Costa Rica). Laszlo is one sophisticated rabble-rouser. A renowned concert pianist earlier in his career, he plays his audience adroitly, making us feel both guilty (for having contributed to the problem) and self-righteous (for being enlightened enough to care) at the same moment.

"Irreversibilities will arrive before we have a chance to act! Civilization is no longer sustainable without a profound cultural mutation. If you think you can live with a significant portion of the world on the edge of survival, you are living on another planet," says Laszlo, who has long argued that a confluence of factors will lead to a collapse of global civilization somewhere between 2012 and 2015.

Aftermath Scenario: Greenland Melts, Northern Hemisphere Sweats

Jakobshavn Isbrae, the fastest-moving glacier in Greenland, which doubled its speed in the first few years of the twenty-first century, doubles speed again, and by the beginning of 2012 the mammoth iceberg maker is scudding along at 15 miles per year, roughly the equivalent of Manhattan sliding over to central New Jersey. Think of an ice cube left in the kitchen sink. At first it won't move at all, then it will begin to drift. Soon enough, there comes a moment when that cube, or what's left of it, will dart in whatever direction the puddle it created decided to flow. For Jakobshavn, that moment comes in 2012.

Jakobshavn is below sea level, meaning that when it melts, warm ocean waters rush in upon it, accelerating the decrystallization from ice to liquid, the way a trickle of tap water would hasten the demise of our hypothetical ice cube. The meltdown into the North Atlantic makes for some excellent video footage and computer graphics, and prompts stern speeches about global warming and rising sea levels. Commanders of the U.S. military bases there confess that they had long been aware of the problem but were forbidden to disclose this information, ostensibly for national security reasons. Native Greenlanders, who for the most part have been happy with the warming of their homeland since it enabled them to grow cabbages and other crops they were formerly forced to import, now travel the world to tell of the swamping of their homeland. But the plight of the Nordic indigenous proves less motivating than these brutal, simple facts: approximately one-quarter of the world's population, about 1.5 billion people, lives within 100 kilometers of the coast. Ten percent, or 600 million, lives at sea levels of 10 meters or less.

The largest metropolitan area directly in line to suffer inundation from Greenland's glacial meltdown is New York City, a surly veteran of catastrophe if ever a city was. Much ado, drama, and malarkey, but very little engineering or construction work, accompany calls to build the mother of all sea walls to protect Wall Street, the United Nations, the Statue of Liberty, and the city's other crown jewels. Fresh Kills, Staten Island, the largest landfill in the world, is eyed as a raw material for building landfill levees, and for a few brief months, trash becomes gold. There is even talk of conscripting regional residents for the massive crash-construction project, but the whole project is strangled in red tape and quibbles about centimeters.

Tokyo, Shanghai, and Mumbai, also lying no more than 10 meters above sea level, go through similar struggles. Although Paris, London, and other cities of Western Europe lie a few meters higher, their real threat, given their northerly latitudes, is the growing possibility that the global warming causing Greenland's meltdown will in fact snowball into a new Little Ice Age, as frigid melt water pours into the ocean off Europe's northwestern coast, raising sea levels and depressing temperatures throughout the densely populated and economically and culturally advanced region.

The polar ice cap melts faster than reports can be issued. Melting begets melting, as greater and greater surface areas of ice are exposed to the warming ocean water. As the ice, which is very efficient at reflecting sunlight back into the atmosphere, thus cooling surface temperatures, is replaced by water, which absorbs heat and light, the Earth's albedo (literally, whiteness, or degree of reflectivity) diminishes, thus accelerating the warming process and filling the air with unprecedented amounts of moisture. All of this newly available moisture makes for unprecedented storms, including blizzards, which pummel Western Europe. Hopes that the new snowfall will replenish the ice are dashed as the new snow cover simply melts right along with the old.

Crops in Europe falter, touching an ancient nerve. Although declines in food production are slight, panic is outsize as ancestral memories of famine are recovered. In this supercharged atmosphere, what would normally have been nothing more than a human interest story kicks off a major political ruckus, as a survivalist collective in the Dordogne region of central France attempts to take up residence in the Gouffre de Padirac, the giant, cathedral-like cavern that housed hundreds of French families during the Hundred Years' War, 1337 to 1453, between Britain and France. The suggestion that a century-long conflict might occur in today's oh-so-civilized Europe no longer seems laughable.

The onset of the Little Ice Age in the early 1300s decimated agriculture and led to the Great Famine, a tragedy immortalized in the Hansel and Gretel fairy tale. A new, sensationalist book by a hitherto obscure medieval historian describes in gory detail how, as crops failed and the price of grain soared, the people of the British Isles, northern France, the Benelux countries, Scandinavia, Germany, Poland, and the Baltic states descended into starvation, cannibalism, and plague, which killed up to 10 percent of medieval northern Europe's population. Known as the Black Death, the plague was clearly related to the famine's weakening of people's immune systems and of their desire to live.

Rancid food gobbled up by desperately hungry people also takes its toll, causing diarrhea and other intestinal ailments that lead to demise. Grim jokes circulate about ergotism, known in the Middle Ages as St. Anthony's Fire, a weird, horrible affliction caused by eating moldy grain. Ergot, it turns out, is the raw material from which LSD is made. The ergot fungus attacks its victims' muscles, causing spasms and convulsions. Soon enough, blood is cut off from the extremities, leading to gangrene, amputations, and death, though not before a series of wild hallucinations. The symptoms of ergotism, particularly the hallucinations, inspire contemporary European performance artists to surreal, hilarious heights of bizarre creativity. For the first time in anyone's memory, European popular youth culture proves edgier and more popular than its American or Asian counterparts. Neo-Surrealism is the label stuck on this movement by highbrow critics, but the performers prefer "LSD," for lysergic acid diethylamide, which drug has burst back into popularity, and for the acronym's oblique reference to the need to warm the world back up, as in Lick Santa's Dick.

The Southern Hemisphere, which holds about 30 percent of the world's population, is hammered by the meltdown of Antarctic glaciers. South American agriculture, particularly the fruit and vegetable crops of Chile, is particularly hard hit by the freeze, as are New Zealand and Australia. Pacific island nations such as Fiji and Tahiti struggle vainly against inundation, and sadness pervades the global psyche as such paradises are fouled and sucked under.

A new school of thought downgrading the importance of human beings in the process of history suddenly seems sensible and accurate. Traditionally, history has been understood as the story of people, acting individually or in groups varying in size from small bands to empires and, now, global conglomerates of various ilk. People were seen as the prime movers, as heroes, villains, victims, or all of the above. The most significant exception to this human-centric viewpoint was the orthodox theological perspective, which sees God as the prime mover, with men and women acting out God's will. Now the intellectual world comes, grudgingly, to understand that men and women act out the will of Mother Nature, also known as Gaia. Gaia is a prime mover, a causer of wars, bounty, discovery—a maker, perhaps the most important maker, of history.

The Gaian historical perspective holds that we do what we do primarily because we are responding to natural catastrophes as well as longer-term changes in climate cycles. The Hundred Years' War, for example, was less a set of decisions made by the rulers of the time than an unwitting response to changes in the natural environment known as the Little Ice Age. Same holds for the era immediately preceding the Little Ice Age. The Medieval Warming Period, approximately AD 800 to 1300, saw temperatures in the North Atlantic region warm approximately 1 degree C, or 1.8 degrees F. The seas of the North Atlantic region were therefore ice-free for longer portions of the year, enabling Viking explorers to venture farther afield and eventually discover Vinland, now known as North America. Then came the Little Ice Age in roughly AD 1300, a climatic counterbalance that was also a regional, not global, phenomenon, and the seas were once again choked with ice and the route to Vinland was blocked. Thus it was that the Vikings never truly colonized the land they discovered, an achievement that almost two centuries later would fall to Columbus and his successors, who launched their expeditions from warmer seas to the south.

So who or what is the protagonist of this story? Leifur Eiriksson, the Viking sailor most often credited with being the first to make it to Vinland? Or Gaia, who, albeit unintentionally, warmed the northernmost reaches of the North Atlantic to the point where it became navigable? The Medieval Warming Period provided the opportunity and incentive for Nordic seafarers to enhance their skills and shipbuilding technology, to fulfill their destiny as courageous explorers. Gaia giveth, and Gaia taketh away, this time in the form of a five-centuries-long cooling period that once again made the northern seas impassable.

Is Gaia also a maker of psychology? The big, fat human ego has ever been loath to consider itself subject to, and much less dependent on, the cycles of the natural world, above which we have so impressively risen. We don't normally consider our most intimate thoughts and emotions subject, particularly, to conditions in our natural surroundings. A manic-depressive, for example, would still hit his or her highs and lows no matter whether the weather is warm or cool. Oh, violent storms might spark a few emotional outbursts; a prolonged stretch of balmy weather may take some of the edge off the most intensely manic moments. We grudgingly allow how

Seasonal Affective Disorder (SAD) can make us tend toward melancholy in midwinter, and we might even acknowledge such a thing as Climate Change Affective Disorder (CCAD), in which individuals living much of their lives in the same locale are disoriented by changes in seasonality, but for the most part our working assumption is that the external world is unimportantly related to conditions within our own psyche. That is, until 2012, when conditions in the external world rapidly became more chaotic and extreme, climatically, financially, and in terms of threats to personal security. The rising school of Gaian psychology argues that these external circumstances are ultimately the most important influence on conditions in our internal world, the one we have long prided as uniquely our own.

While the world shivers, the complexities of the Sun-Earth energy relationship are hotly debated. Neo-Copernicans argue that it is the Sun that doth the giving and taking away. Virtually all of the Earth's energy comes originally from the Sun, meaning that any variations in solar output affect our planet, our civilization, and our daily lives. Thus, it may have been that the downtick in temperatures experienced during the Little Ice Age was caused in part or in whole by a cyclical decrease in solar luminosity. The popular conception of the Sun as a great big space heater far, far away warming everything around it proves to be a misleading analogy. The Earth is less like an inanimate object whose temperature rises and falls in direct proportion to the amount of heat energy it receives from the Sun than it is like a money manager skilled at handling variable income streams. Such a money manager would save in the good times and draw on those savings in the lean. The Earth does not "think" the way that a person does, but rather adjusts automatically, an ability that has evolved over 4 billion years, during which time it has become, as Lewis Thomas, the incomparable philosopher of science, once put it, "marvelously skilled in handling the Sun."

Anyone who has ever swum in the ocean on Labor Day weekend can sense firsthand how the global heat management system works. Land temperatures are cooler in September than in June, but the water temperatures are usually warmer, for the simple reason that the ocean has had all summer to heat up. Water, denser than earth, takes longer to warm up and also longer to cool down, essentially acting as a heat bank, with the net effect of modulating seasonal temperature changes, particularly in coastal areas. If there were no such heat management system, water would be at its hottest in late June, when the Sun is strongest, and uncomfortably chilly for that Labor Day swim.

Rainstorms pummel the London Summer Olympics held in August 2012. When pundits dub the Games the Apocalypse Olympics, or even Apocalympics, the International Olympic Committee responds asymmetrically, banning all journalists who dare use the phrase. The IOC even secures the arrest of an irritating American blogging from his apocalympics2012.com website. The blogger's rhetoric, while vulgar and vituperative to the IOC, cannot fairly be construed as a call to violence, but his daily declarations that these are the last Olympics before the Apocalypse are severely disturbing, and do indeed incite panic among some of his followers. High-minded commentators hold forth on the threat to free speech. PEN International mounts a campaign to free the blogger. But sports fans around the world sorely desire the diversion of the Olympics and take satisfaction in having someone to scapegoat for the failure of the Summer Games.

The deepening freeze leads some to seek salvation in far-fetched trillion-dollar technological fixes that would mine magma from the Earth's crust as a source of heat, or that would orbit in space, collecting the Sun's rays and lasing them back to Earth. Others would mine geothermal energy and pump it up to the surface. Nothing ever comes of any of this, because the projected costs are too high, the time lags too great, and the outcome too uncertain.

The failure of science to resolve the situation spurs spiritual hysteria. New Age, pagan, and indigenous nature worshippers rally millions of followers to supplicate the Earth deity. Goddess cults, largely of female adherents, prove remarkably fertile and creative artistically. Mayan shamans, whose ancestors' predictions concerning 12/21/12 as the birth of a new era are now widely seen as apt, are thrust into the limelight, sometimes to the point of being adored as demigods. Some prove equal to the task of retaining their humanity and humility in the face of the explosion of adulation. Most do not.

Ancient suspicions that the Earth has powerful and vital nodal points, akin to the acupressure/acupuncture points along the meridians of the human body, are substantiated scientifically. Perhaps a better understanding of the energy dynamics of these geomagnetic portals can aid in our recovery. For the most part, scientists and policy makers ignore this suggestion as too far-fetched, but it nonetheless becomes a hot political issue as proponents of this newly resuscitated theory present the historical aspect of their argument, that over the millennia, churches have been purposely built on these geological nodal points, presumably in an attempt to subjugate their power and the pagans who would worship it. The Roman Catholic cathedral at Chartres, France, is positively identified as one of the churches thus constructed. Extremists of this geophilic philosophy demand that the structures be destroyed.

The Roman Catholic Church and other mainstream orthodoxies astonish their critics, and perhaps even themselves, with their "guilty with an explanation" response to these accusations. They admit that many churches were built on such nodal points, sometimes by accident and sometimes on purpose, though not so much as to squelch their power as to focus and transmit it to the Lord. Once again, Chartres focuses the debate, but this time as an undeniable example of man-made magnificence springing from the raw force of nature. Churches, largely empty in Europe except as museums, are filled after they are reunderstood as celestial transmitters, places for people to come together and thus amplify their entreaties to the Almighty for an end to the climate madness. The role of priests, rabbis, ministers, and other traditional clergy is to focus and culminate the salvation effort. It helps that wine is served after the services.

A profound rapprochement between science and religion emerges, with the understanding that ACC forecasting is beyond the realm of either discipline alone. The best route to salvation from the current cooling crisis is the respectful collaboration between earth scientists, on the one hand, and, on the other, shamans, sages, and others with exceptional intuitive sensitivity to the ways of the natural world. If nothing else, the two groups compensate for each other's deficiencies. What the computer modelers lack in practical wisdom and feel, the shamans lack in global perspective, shamans generally devoting their lives to and deriving their wisdom from close emotional and spiritual relationships to the locales where they live and serve.

Laypersons subscribe to different theories regarding the cooling trend with the same ardor they might have formerly reserved for politics and religion. But relief organizers see it all as rearranging deck chairs on the *Titanic*. No one, not even the geniuses of the age, wants to admit that the time for understanding is past. Everyone pays lip service to the urgent need for civil defense, but for some strange reason, this lifesaving, civilization-saving necessity never gets the respect or funding or legal empowerment that we so desperately need it to have.

As temperatures plummet mercilessly, demand for heating oil and natural gas spikes and spikes again. North Sea oil production, Europe's ace in the hole, is threatened by increased storm activity, but the crews labor on gamely, the new heroes of the region as they risk their lives to deliver the vital fluids and gas. Ultimately, however, the European Union's economic welfare is in the hands of Russia, the largest petroleum supplier to the region. NATO, the U.S.-dominated security alliance, is torn by East-West squabbles. Cynically, European leaders blame the United States for having caused the global warming mess and for not having curbed greenhouse gas emissions early enough. Oil-rich nations of the Middle East play both ends against the middle, siding publicly with the Europeans while continuing to sell their oil and gas, at skyrocketing prices, to the United States.

The Arabs tighten their alliance against Israel. Israel, feeling the squeeze and terrified of another Holocaust, lashes out genocidally against Palestinians and other persons of Arab heritage living within its perceived domain. Impelled by the November 2012 crash of Noah's comet, Armageddonists of all religious stripes—Muslims, Christians, and Jews—organize for the final, biblical battle.

By mid-2013, the weather begins to clear and temperatures rebound, as does the overall mood of Western society. The air over the Northern Hemisphere, remarkably clean after so many months of rain, fulfills Lovelock's prediction that the atmosphere, devoid of soot from hydrocarbons, will heat that much more quickly. Temperatures spike and spike again. The global warming that caused the glaciers to melt down resumes inexorably, and the next phase of abrupt climate change begins. But there is a blessing in all this: the cooling has bought us time, and has demonstrated to all but the terminally stubborn that climate change is for real, and that we must prepare and adapt, or the next phase will be much, much worse.

6

LIFE WITHOUT CUCUMBERS

I never gave a second thought to cucumbers until one evening in early January 1990, at a New Year's party at the American Embassy in Prague. The communist regime in what was then Czechoslovakia had just fallen, replaced by a pro-Western democratic government headed by Václav Havel, whom *Playboy* magazine had sent me to interview. There was plenty of good liquor, ham, sausage, and cheese on hand, but when the tray of fresh cucumbers was carried out, a thrill rippled though the room and the crowd converged on the waiter. The U.S. ambassador, Shirley Temple Black, who had spent much of the party icily turning down requests to sing "On the Good Ship Lollipop," suddenly found herself deserted for thinly sliced vegetables. (Like tomatoes, cucumbers are botanically classified as fruits but are considered vegetables in culinary terms.) Though I would gladly have foregone my share, my host at the party proudly announced that the cucumbers had been flown in from West Germany just for the occasion, then elbowed us a path through to the tray. So to be polite, I ate a sliver. It was fantastic! Not that that specific cucumber necessarily tasted better than any other, just that eating it at that moment, after having lived for two weeks on beer, butter, ham, and cheese, I could really appreciate this cool, crisp, subtly sweet gift from the Earth.

AFTERMATH

Fresh cucumbers in the middle of winter got people to talking about the good life, and each generation had its own particular take. The youngest guests, in their twenties and thirties, rarely if ever had produce in the winter, so for them the cucumbers presented a clear-cut choice: stay in Czechoslovakia and wait for prosperity to come to them, or, if they could manage it, migrate someplace where there was fresh food aplenty. Marina, a university student and competition tennis player who worked afternoons cleaning the streets, was particularly eager to know about supermarkets—who was permitted to visit them, what foods they had inside—so I regaled her with tales of pineapples in winter, the fresh deli counter, and double-coupon Wednesdays.

Hard experience had taught the eldest guests that when good food comes their way, eat first, talk later. Why ladle gray-matter meaning on such a delicious and unexpected midwinter treat? Although I didn't put it together at the time, those in their seventies and eighties had, in their youth, suffered the fallout from what still stands as the single worst famine in history, known as the Holodomor, perpetrated in 1932–1933 by the megalomaniacal Soviet dictator Josef Stalin against the people of Ukraine, which bordered Czechoslovakia to the east. All told, roughly one-quarter of Ukraine's population, about 9 million people, perished, mostly by starvation, an excruciating death that usually takes weeks, as the body literally attempts to eat itself to stay alive. In Stalin's quest to break Ukraine's national identity, his troops confiscated peasants' farms and the grain they had grown and then made it illegal for the farmers to eat any of it. Stealing even an ear of corn was punishable by ten years in prison, where the prisoner would probably die anyway. Approximately 1.7 million tons of grain, or about 400 pounds per person who starved to death, was exported by the Soviets to the West during this period, the depth of the Great Depression. (To provide a measure of spiritual protection against famine and other such cataclysmic horrors befalling us in 2012 and beyond, I propose a simple ceremony, known as the Circle of Mud, which I'll explain in the Epilogue of this book.)

The cucumbers threw some of the middle-aged partygoers into a quandary. Though grateful for the midwinter treat, they seemed bitter about having been forced to spend such a large chunk of their lives without the simple goodness nature had to offer. These folks were old enough to remember life in the decades following the Second World War, when Czechoslovakia was dominated by Soviet communism, an increasingly intolerable situation that led in 1968 to the Prague Spring, a budding democracy movement, whose slogan was "Socialism with a human face." Soviet tanks quickly crushed the rebellion, and there followed a vengeful crackdown, after which basic freedoms of speech, assembly, and travel were revoked. The material quality of life plummeted, the Russian government charging Czechoslovakia dearly for the expense of the invasion and for maintaining occupying forces in their land. During a visit to the Agricultural Institute in Nitra, in what is now Slovakia, I learned that, post-1968, agriculture was reorganized to fit Marxist-Leninist ideology rather than the natural growing cycles of the land. No force, not even Nature herself, was deemed a match for the will of the proletariat. Farmers were forced to graze their cows in grassless fields and plant corn, a flatland crop, in the hills. Because that was the Plan. Instead of grain, the Communist Party elite (who privately procured whatever delicacies they wanted) ground down their workers' souls.

When I expressed my sympathy, Olav, a literary translator in his fifties, cut me off cold. He was ashamed that his generation had been the ones to blow it. His anger was omnidirectional, focused both on the Soviet bullies and also on the democratist rebels who had upset the apple cart. He was embarrassed at how far living standards had fallen, that basic foods such as fresh cucumbers had been lost on his generation's watch. He was bitter that he had been put in a position of having to be thankful for the return of that which, by all rights, should have been his all along.

A lack of cucumbers, of course, does not an apocalypse make, but still, how radically would the quality of our lives have to diminish in order to be as grateful for fresh vegetables as the folks at that embassy party were? For us, who have grown up with so much, Olav's embarrassment and bitterness is an object lesson we had better start learning now. Even a modest decline in living standards will be profoundly disturbing, not so much because of the physical lack of fresh vegetables or whatever, but because it undermines our treasured assumptions about the future. Who wants the glare of disapproval from the kids, the disrespect that will inevitably be directed toward us as it becomes evident that we are the first generation to fail to uphold the lifestyle status quo. Of course, had we risen, rather than fallen, to that same level of dietary affluence—had we gone from a position of never having had any fresh cucumbers at least to having them in season—our outlook on life would be upbeat and optimistic. Same net yield of fresh vegetables, completely opposite human results.

SNAPPING THE FOOD CHAIN

What happened to the Czech and Ukrainian farmers under communism is, in effect, what is happening to Western farm workers today—not the human ones but the insect workers, particularly honeybees, who pollinate from one-quarter to one-third of all the produce we eat. Some ninety of the world's most common food crops, including apples, oranges, avocados, tomatoes, spinach, lettuce, squash, and, of course, cucumbers, depend on honeybees for their survival, as do important feed crops such as alfalfa and red clover. Whatever the intended harvest, the process is pretty much the same. Each year, around the time when the plants flower, hives are carried into the growing fields and evenly distributed around the edge. The bees, whose legs have a static electricity charge that picks up pollen, are then released to go gad about from blossom to blossom to do their life-giving thing. Except now the busy little honeybees are dying off in droves.

Colony Collapse Disorder (CCD) has caused hundreds of thousands of honeybee hives around the world to fall lifeless. First reported by beekeepers in 2004 but reaching critical momentum in 2006, CCD has affected 50 to 90 percent of all commercial bee colonies in the United States, according to a BBC news report. For honeybees, CCD seems the perfect storm of evil, a combination of parasites, viruses, and insecticides. Of the various parasites attacking honeybees, mites, essentially tiny spiders, are the most lethal: the varroa mite from Asia attaches itself to the outside of the bee and sucks its fluids dry, while tracheal mites from South America infest the windpipe and slowly choke their hosts to death.

Israeli Acute Paralysis Virus (IAPV) is found in virtually all collapsed hives but in very few healthy ones. Whether or not IAPV is a cause of CCD, or simply a marker, is not yet known, but pray that it's just a marker, because it's highly unlikely that scientists will come up with a treatment for viruses in bees, according to Dr. Jeff Pettis, head of the United States government's Bee Research Laboratory. Worse yet would be that CCD is caused by a cocktail of viruses, virtually impossible to inoculate against. Working with researchers at the United States Department of Agriculture, University of Illinois entomologist May Berenbaum analyzed the genetic transcripts of thousands of bees from around the United States, those of healthy bees as well as those of bees stricken with CCD. They found that the stricken bees could not manufacture the proteins necessary to defend themselves from multiple viruses simultaneously attacking and overwhelming their immunological machinery, as reported in Proceedings of the National Academy of Sciences (August 24, 2009).

Intentionally or not, mankind's accelerating emissions of pollution and destruction of natural habitat may well be weakening the food chain and helping to kill off the bees. Although Berenbaum's team found no indication of elevated expression of pesticide response genes in bees stricken with CCD, other researchers are not so sure that overexposure to pesticides has not contributed to the rise of this "cocktail of viruses" to which hives are proving so susceptible. IMD (imidacloprid), a relatively new insecticide whose use has quickly become widespread in the United States, may well bear some blame for CCD, according to Michael Schacker in A Spring Without Bees: How Colony Collapse Disorder Has Endangered Our Food Supply. IMD reduces honeybees' desire to feed, starving out the hive.

Schacker's intellectual forebear is Rachel Carson, whose legendary book, Silent Spring, spurred the international movement to ban the insecticide DDT; her work is proof positive that pernicious chemicals—compounds synthesized by human beings and applied to specific purposes, such as the elimination of pests—can indeed have malignant ecological ramifications far beyond their intended use. So can the arbitrary elimination of said chemicals. When pests run rampant, human health suffers by the resulting propagation of disease, and also, of course, by losses to the food supply. However, the fact that IMD seems pernicious to the honeybees, on which we depend so utterly, argues strongly for increased control of this dangerous compound.

Contaminants in the high-fructose corn syrup that commercial beekeepers often use to feed their hives have also been cited as a possible contributor to CCD. So has microwave radiation emitted from cell phones, believed to interfere with bees' internal navigation systems and thus causing their death by preventing their safe return to their hives. This would certainly help explain the bizarre phenomenon that the CCD-stricken hives are not only lifeless, but often ghost towns physically devoid of bees.

"America's honeybees are sitting ducks," writes Jeffrey A. Lockwood in Six-Legged Soldiers: Using Insects as Weapons of War. Lockwood explores the honeybees' vulnerability to bioterrorist assault. He cites a 1998 Washington, D.C., workshop on agricultural security that found our nation's bees to be "an inviting and unappreciated target." Unguarded apiaries could easily be infiltrated by an enemy, spreading spores, fungi, viruses, or mites such as the ones described above. Could terrorists therefore have caused CCD? According to "Countering Agricultural Bioterrorism," a report that the United States Department of Agriculture commissioned the National Research Council of the National Academy of Sciences to prepare in 2000, national, state, and local mechanisms are insufficient "for effectively deterring, preventing, detecting, responding to and recovering from agricultural threats." The study's major recommendation, for the nation to implement a "bioterrorism rapid response strategy," was apparently never implemented, soon overshadowed by the more pressing concerns of 9/11 and the Iraq and Afghanistan wars.

If the honeybees don't do their pollination work, who or what will?

"Each female [cucumber] blossom is only receptive to being pollinated for one day, and each blossom requires an average of eleven bee visits to set up a well-shaped cucumber," according to John T. Ambrose, a North Carolina State apiculturist. Agricultural researchers have begun looking for other pollinators to take over the work from the honeybees, but eleven times in one day is a hard performance to match. Bumblebees are prime candidates; however, they have the physical skills but not the work ethic, particularly for demanding partners like cucumbers. After three or four goes, the big bumblebees are spent.

AFTERMATH

Geneticists might be able to reengineer honeybees to withstand the perils of CCD, though even if this could be accomplished, we might have some qualms about releasing into the environment millions of mutant bees. Alternatively, the nanotechnologists might try their hand at coming to the rescue by building "beebots," tiny robotic bees that could replicate the work of their natural counterparts, maybe even without the stingers. It's a long shot, though, and precautions would have to be taken against equipping the beebots with the ability, say, to receive signals and swarm at the command of whoever was controlling the software.

"Most important of all, is there a way to avoid this 'Beepocalypse?'" asks Schacker, who demands that we go organic, reducing the use of chemical pesticides, particularly IMD, and replacing them with natural compounds less harmful to the insects. Since bees have keen senses of color and fragrance, farmers could plant a variety of flowers sequenced to blossom at different times of the year and in different colors, to create a more bee-friendly environment. Home gardeners should do the same, since the bees their gardens attract can be collected by beekeepers and used in the fields. Pollen-free plants, though conveniently hypoallergenic, trick honeybees into thinking that there's pollen inside them when there isn't, so they should be replaced by plants that make pollen, which is what bees eat.

All good suggestions, so why aren't there are more names on the "Save the Bees!" sign-up sheet? Bees are not whales. Nobody identifies with them, dialogues with them, or considers them a delicacy. Almost everyone has been stung by a bee, and if they go extinct, precious few of us will mourn their loss, until we realize, too late, what that loss really means. Never mind that Albert Einstein never actually said civilization would collapse five years after the bees disappear. That was a false attribution made in 1995 by striking Belgian beekeepers trying to drive home the point of how utterly dependent we humans are on bees for our agriculture and sustenance. Of course, just because Einstein didn't say it doesn't mean it's wrong. Five years from 2006, when the bees began disappearing en masse, would put global famine right at the doorstep of 2012.

Grand and glorious human civilization brought down onto its bloated bellies by a lack of buzzing, stinging insects? How pathetic would that be?

"Are the honeybees trying to tell us something? Are they the canary in the gold mine, warning us of hidden changes to the planet and all of humanity? On a deeper level, are the bees telling us we are unaware of a deep systemic problem threatening our own species, are we missing the big picture here? Could our own human colony become undone, through some kind of 'Civilization Collapse Disorder?' " writes Schacker.

Could just be that the bees' time is up, evolution-wise. That corrosive CCD virus/parasite/pesticide combination might just be the current means to their inevitable end. Perhaps this is just the latest stage in a cycle of death and rebirth played out again and again throughout the evolution of all species. But there's nothing "c'est la vie" about the honeybees disappearing. Schacker points out that over the past 100 million years, flowering plants have coevolved closely with bees—neither would exist nearly as plentifully without the other. Most fruits and vegetables would never have come to be had not bees been around to do the hard work of pollinating them. Of course, if bees and flowering plants do indeed go down the evolutionary drain together, some new and more fruitful partnership will undoubtedly emerge, eventually. Too bad that such processes normally take thousands of years.

Odds are good, because God is good, that the busy little honeybees will fight their way through CCD and continue on down their 100-million-year evolutionary path. Some combination of human intervention and natural mutation favoring those bees who can resist viruses, mites, chemical toxins such as IMD, and the other threats to the health of the hives, will, knock on wood, yield a rising generation of CCD-resistant bees. But God is no chump, either. Even if the honeybees do manage to return this time, their disappearance, coming on the heels of an earlier unexpected drop in their population in the early 1990s, may well represent a progressive weakening of this vital link in our food chain.

You don't have to be religious or superstitious to heed as a warning sign the fact that some of the hardest-working and most ecologically beneficial creatures are vanishing from the planet.

THAT UNEASY FEELING . . .

Am I the only one who gets an uneasy feeling about this, or is it more than just coincidence that the butterfly, amphibian, and bat populations are also in trouble, along with the honeybees? Butterflies are vanishing at a truly alarming rate, according to one of the most exhaustive and pro-longed nature studies ever conducted. Every twenty years, beginning in the mid-1940s, a team of up to twenty thousand naturalists inspects the entire British landscape to compile an atlas of birds, butterflies, and wildflowers. The most recent atlas, published in 2004 and reported on in *Science* magazine, found that nearly one-third of native British wild plant species, one-half of native British bird species, and nearly three-quarters (71 percent) of native British butterfly species have fallen in numbers over the past twenty years.

"The ploughing of heathland and the draining of wetlands have resulted in complete destruction of some [butterfly] habitats, while others have become degraded as a result of other human activity, such as pollution," concludes Jeremy Thomas, a researcher at the Centre for Ecology & Hydrology in Dorset, England, who led the study.

Although butterflies are adroit pollinators, particularly of flowers with fragrances too faint to draw bees or other insects, the greatest tragedy would, to my mind, be the loss of the butterflies' ineffable beauty. Imagine sitting down with your grandchildren to watch a video about butterflies, telling them what it felt like to have one alight on your wrist, and then having to explain why they will never have a chance to experience that pleasure.

"The results are appalling," says Thomas of the butterflies' disappearance. "This adds enormous strength to the hypothesis that the world is approaching its sixth major extinction event." He explains that the last mass extinction of comparable magnitude is the Cretaceous-Tertiary, infamous for extinguishing the dinosaurs and 70 percent of all other species some 65 million years ago. At least the dinosaurs had an excuse, namely, the giant asteroid or comet that caused their mass extinction. History may well conclude that we were even stupider than those giant, pea-brained lizards, having helped bring extinction upon ourselves.

Cold comfort that frogs, salamanders, newts, and other amphibians will probably precede us down the evolutionary drain. Living, as they do, both on land and in water, amphibians are doubly vulnerable to ecological threats such as pollution and warming. "As the Earth warms, many species are likely to disappear, often because of changing disease dynamics. Here we show that a recent mass extinction [of amphibians] associated with pathogen outbreaks is tied to global warming," explains J. Alan Pounds and thirteen other scientists from three continents in "Widespread Amphibian Extinctions from Epidemic Disease Driven by Global Warming" (*Nature*, January 12, 2006).

Having caught dozens of frogs and toads as a boy, I could wax indignant about the spiritual poverty of life without them, about growing up without having any earthly idea of what Mark Twain was writing about in "The Celebrated Jumping Frog of Calaveras County," his 1867 short story about Dan'l, the champion jumper who couldn't jump because his competitor's owner had sneakily poured buckshot down old Dan'l's throat. Then there's the fact that amphibians, uniquely positioned between water and land, have yielded to medical science a bonanza of valuable novel compounds, including epibatidine, 200 times more powerful than morphine, and the secretions of the dumpy tree frog, which make for an excellent organic mosquito repellent. Frogs and their tadpoles are also an important food source for freshwater fish. But to my way of thinking, the most vital role played by amphibians is that they consume flies, mosquitoes, and other insects by the megaton.

So do bats, which eat up to three thousand mosquitoes per night.

"White-nose syndrome is a fungal infection that has killed some 75% of some bat populations in Massachusetts, Vermont, New York, and Connecticut since it was first discovered in a cave in upstate New York in 2006," according to The Scientist, a respected Internet journal of the life sciences. The white-nose fungus tends to blossom while bats hibernate during the winter, causing bats to burn too quickly through their store of fat, thus forcing them to go foraging in the cold for insects before there are any to be had. Similar die-off rates among these nocturnal flying animals have been reported elsewhere in the United States, Britain, and Europe, for reasons ranging from white-nose syndrome to loss of habitat, particularly the destruction of their caves. Few of us can muster the emotional wherewithal to mourn the disappearance of flying rodents, what with Dracula and all, though a couple dozen more mosquito bites each week could prove enlightening. So might the fact that bats eat a wide range of insects that would otherwise eat plants such as apples, wheat, and cucumbers.

Here's the question not to ask: What will life be like if the honeybees, butterflies, bats, and frogs go extinct? Answering that one could keep a hundred scientists in government grants for a decade. The cascade of possibilities involving what goes extinct where and when, and how other insects, birds, lizards, and mammals might take over the vanishing species' ecological roles, truly approaches the infinite. A dozen super-computers would generate computer models by the terabyte, and the results would still somehow turn out to be uncertain enough to warrant a whole new round of government grants.

Here's the question we need to ask: What should we do to protect ourselves in case the food chain collapses? Common sense says we are headed for a jarring degradation of the agricultural landscape. Reversing that process will take lots of effort and investment, essentially a changeover to an organic-style agricultural system, as Schacker recommends. No arguments here, but that's probably not going to happen quickly enough to repair all the weak links in the food chain, certainly not by 2012.

THE FARMERS' MARKET FUTURE

Most of us tacitly assume the progression from humble plywood farm stands to gleaming supermarkets, with their backlit display cases and automatic misters, to be just that, progress, heading to the future, the direction things, for better and occasionally for worse, are bound to take. Having written, in an earlier stage of my career, the page equivalent of a dozen books about the supermarket industry, having pored over thousands of pages of statistics on supermarket sales, growth, factors in future growth, brand shares, distribution trends, and demographics, I am constitutionally incapable of believing that the whole damn food distribution system could be brought down by the lack of bees. Not after celebrating the day that my report "The Mexican Foods Market" was quoted in *The Wall Street Journal*: "Salsa sales surpass ketchup, signaling demographic shift." No one ever mentioned anything about honeybees, I can guarantee that. But the fact of the matter is that you need tomatoes to make both salsa and ketchup, and you need honeybees to make the tomatoes.

If the honeybees disappear, so will the supermarkets, at least as we know them. Yields of fresh produce, and therefore of all those foods prepared from them, will not be nearly bountiful enough to supply the current retail system. Without the honeybees or some as yet unknown pollinating equivalent, the "supermarket future" could well give way to what I've come to think of as a "farmers' market future," a radically downsized and localized distribution scheme in which a much greater share of our food is grown and consumed locally.

Greenhouses are an excellent defense against the possibility of tomato, cucumber, and other fruit and vegetable crops failing due to the die-off of natural pollinators. Plants grown in greenhouses generally do not require insect pollination; instead, blowers circulate the pollen. Greenhouse fruits and vegetables are therefore generally seedless because fertilization by insect pollination is what creates the seeds. Being seedless, these plants cannot reproduce, so the greenhouses must be periodically resupplied with insect pollinated, field-grown plants, so some honeybees or suitable substitutes are required, but not nearly so many as if the total crop were field-grown. Of course, our taste in cucumbers will have to shift, from field crops such as the big green slicers and kirbies for pickling to seedless varieties known as "burpless" (because the seeds are what make one burp) and wild cukes also known, perhaps too poetically, as "manroot," to slender Persians, delicate Japanese, maybe even that new "c-thru-cucumber," a greenhouse variety developed with a skin so thin that it does not require peeling.

Building a network of greenhouses would have the added advantage of enhancing our security, defending not only against potential food chain collapse due to ecological mishaps but also against possible disruptions in the agricultural import stream caused by war, terrorism, fuel shortages, and/or natural catastrophes. In times of political and economic crisis, it is critical to reduce the distance and travel times between consumers and the vital goods and services they require. This means favoring locally grown produce over that grown in other parts of the country, or abroad. Would this policy constitute "protectionism," that favorite epithet of the free traders? Perhaps, and there's nothing wrong, by the way, with adding an extra layer of protection to society's supply of basic goods and services, particularly perishable foods. Indeed, small, local solar-heated greenhouses can be erected quickly and inexpensively in communities throughout the United States, and pretty much everywhere else in the world.

Sadly, greenhouse construction does not address all of our agricultural needs. Staple grains such as wheat, corn, barley, and rice are strictly field crops. Fortunately, grains do not require insect pollination and are not currently in danger of faltering production. Orchard-grown fruits such as cherries, apples, oranges, peaches, and plums that require insect pollination do not lend themselves to greenhouse cultivation. They are, however, generally amenable to being grown in vest-pocket farms. According to Ambrose, the apiculturist from North Carolina State, small farms of three acres or less do not require the injection of honeybee hives if they are set near wilderness areas from which they can attract enough wild bees to pollinate the crops, assuming, of course, that the wild bees have not succumbed to CCD as well.

Pickup trucks full of produce soldiering from local farms and greenhouses into town all sounds very lovely and bioregional. But farmers' markets are charming only as long as they are picturesque alternatives to supermarkets, not their replacement. While at a certain level we can all appreciate that the intimacy and freshness of a farmers' market is a better way of life than Costco's extrawide shopping carts filled with jumbo multipacks of frozen corn dogs, what a come-down it would be! Zero packaging, as you hand the farmer money and he hands you lettuce. Can the leathery old woman missing several teeth and half an eye still be trusted to know her wild mushrooms well enough to distinguish unerringly between delectable and poisonous? Probably. At Whole Foods you'd pay three times as much for the mushrooms, though the thought of dropping dead after eating them would never cross your mind.

Spiritually, it's a good thing to be thankful for simple pleasures. The Maya believe that when gratitude for the necessities of life—clean water and air, healthful food, sturdy shelter—becomes the prevailing attitude, rather than the exception to the rule, this will be a sign that the new and enlightened post-2012 era is here.

But how to show our gratitude for, say, the humble cucumber? Although I traveled to Russia three times to research 2012, I never made it to Vladimir, a smallish city in south central Russia, to attend the annual Cucumber Festival sponsored by the Museum of Wooden Architecture and Peasant Life. Prizes are given to the farmer who grows the longest cucumber and also to the cook who makes the saltiest pickle. Their parade, as I understand it, is usually led by a twelve-year-old dressed up as the Cucumber Fairy. Our own Cucumber Parade will of course be much grander. Giant cucumber balloons sponsored by local greenhouses will fill the sky and tiny yellow cucumber flowers will festoon our floats. A squadron of unemployed beekeepers, veils down, hive-smokers blasting, will inter their hives around the monument to the fallen honeybee. Mourners of butterflies, bats, frogs, and whatever other species lamented by then will of course be welcome to make their own obsequies.

Okay, so the Rose Parade it will not be. The point is that we can survive the loss of cucumbers and other fresh fruits and vegetables, just as the folks in Eastern Europe did for so long. If the honeybees disappear, we will find new pollinators and/or increase production of these basic foods in greenhouses and other venues that do not require the plants to be pollinated. If other useful insect species are lost due to environmental degradation, we will suffer, but not perish, and in all likelihood our scientists will create workable substitutes for the services these insects had once performed.

What we cannot count on doing is to compensate for such lack of vital biodiversity and for the consequent losses to our health and welfare while at the same time struggling to survive economic disaster and the social chaos that could well result. As it happens, we will reach a destiny-altering turning point in 2012, and it has nothing to do with Mayan prophecy, or with the solar climax expected to occur in that year. Rather, it regards the Kyoto Protocol, which is intended to curb greenhouse gas emissions and combat global warming, and expires in 2012.

CHANGING CLIMATE CHANGE

quitting smoking is a good thing to do, though not necessarily when one is about to get shot in the head or run over by a truck. Same holds for curbing greenhouse gas emissions—it's a good thing to do, but there are probably more urgent threats.

I feel like such a traitor for writing that.

Climate change has dethroned nuclear holocaust, which reigned from Hiroshima until the end of the Cold War, as the sum of all our fears. The difference between the two specters is that safeguarding the world from nuclear destruction was largely in the purview of a tight circle of diplomats and military men, while global warming is something for which everyone is responsible, and which everyone can help prevent. Without a shadow of a doubt, it is wonderful that people and their societies around the world are coming together to minimize waste and pollution from hydrocarbons and other sources, to husband valuable resources such as fresh water and topsoil, to harness solar and wind power, and to protect key ecological areas such as the rain forests, wetlands, and coastal areas from degradation and destruction. Sacrificing frippery and excess for the sake of the common good, much the way that folks served a higher purpose during the Great Depression and the Second World War, is a noble, necessary endeavor. The green revolution uplifts everyone, and the movement to prevent global warming is the engine that drives the green revolution. To do or say anything to undermine this historic spirit of togetherness by appealing instead to base self-interest would indeed be mean-spirited.

However, it could be that we have become a bit fixated, . . .

"Global warming commandeers a disproportionate fraction of the attention given to global risks . . . we may become unduly fixated on the one or two dangers [such as global warming] that have captured the popular or expert imagination of the day, while neglecting other risks that are more severe or more amenable to mitigation," write Nick Bostrom and Milan M. Cirkovic in *Global Catastrophic Risks* (Oxford University Press, 2008).

Neither Bostrom nor Cirkovic are climate change naysayers. Rather, as the principal organizers of a 2008 Oxford University conference on the potential for global-scale disasters, they voice the growing consensus of the current generation of catastrophe theorists who maintain that lesser-understood threats, such as those to the electrical grid and the satellite system, and even the age-old fear of the Earth being hit by a comet, receive less attention and funding in proportion to the danger they pose, particularly compared with the massive global campaign to combat climate change.

This is certainly true. While it is impossible to ascertain the total budget worldwide for climate change research, the number clearly runs into the tens of billions. According to the American Association for the Advancement of Science, approximately $5.6 billion in federal funds were appropriated in 2008 for climate change science and technology. (This was under the ecologically apathetic Bush administration; in all likelihood, this figure will soar under President Obama's stewardship.) A single percentage point of the federal budget would probably be enough, for example, to protect the power grid from space weather blasts and hacker intrusions.

Maybe Bostrom, Cirkovic, and I are just jealous of the support garnered by the climate change crusade. People have always complained about the weather, so naturally the climate change movement has struck a responsive chord. Don't have to be a supernerd to grasp the idea that the Earth is running a fever. But the fact that this native impulse has been painstakingly transformed into a full legal and economic framework for global cooperation is what's truly impressive. That the world has come together to forge the legally binding Kyoto Protocol to the United Nations Framework Convention on Climate Change is an amazing culmination of a global environmental movement among governments and organizations that started with the United Nations Conference on the Human Environment, held in Stockholm, Sweden, in 1972, dramatically accelerated with the United Nations Conference on Environment and Development, held in Rio de Janeiro in 1992, and culminated in Kyoto, Japan, in 1997, with the drafting of the agreement now known as the Kyoto Accord.

The Kyoto Accord commits participants to an average 5.2 percent reduction in greenhouse gas (GHG) emissions, including carbon dioxide, methane, and fluorocarbons (ozone-layer depleters), below 1990 emission levels. The underlying political philosophy of the agreement was that developing nations such as China and India were exempted from numerical limitations because they had not in the past caused nearly as much GHG emissions as developed nations had. The United States was therefore asked to commit to a 7 percent reduction and the European Union to an 8 percent reduction. The Kyoto Accord was ratified by 183 nations, though not by the United States, which until recently was the world's largest greenhouse gas emitter. It has since been surpassed by China, although on a per capita basis, U.S. emissions outstrip China's by a factor of 4:1. Over the period 1992 to 2007, China and India have taken rich advantage of their exemption from numerical limitations in the

Kyoto Accord by more than doubling their GHG emissions. U.S. emissions have risen about 20 percent over that same period, with Western Europe achieving a small net decline.

The carbon-trading market is the financial cornerstone of the new "green economy" blueprinted in Kyoto. In what is also known as the "cap and trade" system, governments and major corporations are annually allotted "caps," set amounts of greenhouse gas they are permitted to emit. If they emit less than the caps specify, they may sell or trade most of the remainder amount to other entities who desire to pollute more than their own caps allow. Polluters who wish to exceed their caps may also purchase credits from developing nations who, in selling such credits, forgo their right to exploit their own natural resources. Although this system demonstrably helps to reduce carbon dioxide levels in the atmosphere, it is highly vulnerable to rhetorical slurs; extremist Greens skewer the "right to pollute" as moral hogwash, demanding to know who sets the pollution caps and if they are too high. At present, the "cap and trade" system flourishes in Western Europe, where Kyoto emissions limits have been adopted as law, and has begun to perk up in the United States, where compliance, though currently voluntary, is expected to become mandatory under the Obama administration. The economic potential is staggering: in 2007, *The New York Times* predicted that the carbon-trading market would become the largest business in the world.

The Kyoto Accord provides for nonpolluting energy sources such as solar and wind to garner carbon credits in proportion to how much carbon pollution would otherwise have been emitted had the energy been generated by burning fossil fuels such as oil or coal. Thus, a kilowatt of wind energy generates revenue not only from the sale of the energy, but also from the carbon credits that take into account the comparative cleanliness of this sustainable form of energy production. Carbon credits can also be generated by reforesting decimated lands, thereby providing trees and other foliage that absorb carbon dioxide from the atmosphere. More complex is the doctrine of "avoided deforestation," whereby landowners in developing nations are paid not to cut down their forests. This preserves the natural environment, prevents the massive emissions of carbon dioxide that would have occurred had the land been clear-cut or, worse, burned to create fields, and enables the trees to continue doing their job of absorbing greenhouse gases. (In the interest of full disclosure, I am the founder of Treevestors, Inc., a tiny start-up carbon-credit brokerage specializing in forestry projects.)

The first commitment period of the Kyoto Accord runs from January 1, 2008, through December 31, 2012, during which time the prescribed GHG emissions reductions are supposed to be achieved. At present, it appears that the Kyoto targets will not be met, barring a catastrophic global economic depression that severely retards manufacturing and industry. In fact, overall GHG emissions are headed for an increase over 1990 levels, though certainly not as much as they otherwise would have increased without the accord. It is interesting to note that the greatest reduction in GHG emissions had nothing to do with Kyoto, but rather has come with the fall in the early 1990s of coal-addicted communist economies, the most reckless polluters in history. Advanced industrialized nations, categorized as "Annex 1" nations in the Kyoto jargon, will face stiff penalties for failing to hit their GHG reduction targets: an additional 30 percent reduction over and above the reduction to which they had previously committed (and then reneged on) and possible suspension from the international carbon-trading market created by the Kyoto Accord. Of the Annex 1 group, only Sweden and the United Kingdom are on track to meet their reduction targets. Whether or not the nations hit with Kyoto penalties will actually make good on what they owe remains to be seen, though the odds of anything near full compliance are quite low.

AFTERMATH

In December 2009, the United Nations Framework Convention on Climate Change reconvened in Copenhagen, Denmark. Designated representatives from 150 nations emitted clouds of carbon dioxide as they negotiated the next phase of climate change legislation. It is still difficult, several months out, to discern the true impact of that at times riotous gathering. Suffice it to say that the meeting should be considered a success simply for having been held, right at the start of the flu season, with the swine flu epidemic raging as it was that year.

KYOTO WHO?

Now that would be ironic: high-minded men and women from almost every point on the globe gathering in the midst of flu season in damp and chilly Denmark, where they pass around the swine flu virus as they orate, huddle, and caucus about saving the planet, then return back home to spread the pandemic. Since some folks believe that imagining something "gives energy" to it, thereby increasing the likelihood that the envisioned outcome will actually occur, I will dwell no more on this tragic scenario. It is kind of weird, though, stepping back for a moment, to think that right at the start of the 2009–2010 flu season, in the midst of what the WHO warned could well become a pandemic infecting 2 billion people worldwide, environmental health representatives from around the world met to discuss not the pandemic but what is still a largely theoretical threat of global warming that may or may not impact us decades down the road.

Perhaps the more constructive way of looking at it is, if the world can come together to create laws and a marketplace to combat climate change, we can damn well do the same to help prevent the spread of infectious diseases, which, all told, account for approximately 25 percent of the world's deaths each year. The transmission of viruses, person to person, animal to animal, and animal to person, is beyond any shadow of a doubt a grave threat to human health. For example, the HIV virus that causes AIDS has killed more victims, shaken more societies, orphaned more children, and destroyed more hopes than carbon dioxide has, Lord knows. Bostrom and Cirkovic write:

> Pandemic disease is indisputably one of the biggest global catastrophic risks facing the world today, but it is not always accorded its due recognition. For example, in most people's mental representation of the world, the influenza pandemic of 1918–1919 is almost completely overshadowed by the concomitant World War I. Yet although WWI is estimated to have directly caused about 10 million military and 9 million civilian fatalities, the Spanish flu is believed to have killed at least 20–50 million people. The relatively low dread factor associated with this pandemic might be partly due to the fact that only approximately 2–3% of those who got sick died from the disease. (The death count is vast because a large percentage of the world population was infected.)"

It is sobering to note that had the latest swine flu epidemic been accompanied by that same 2 to 3 percent mortality rate, the death toll would be in the 40 to 60 million range, given WHO's estimate that a full-blown pandemic would infect approximately 2 billion people. God forbid the H1N1 swine flu virus one day mutates into something more lethal, perhaps by genetic combination with the far more malevolent, though less contagious, H5N1 SARS virus.

"If that happens, I will retire immediately and lock myself in the P3 [ultra-secure] lab. H5N1 kills half the people it infects. Even if you inject yourself with a vaccine, it may be too late. Maybe in just a couple hours it takes your life," says Dr. Guan Yi, the courageous Hong Kong virologist cited by *Time* magazine for having stopped the deadly SARS.

epidemic cold in its tracks by insisting that Chinese authorities allow the slaughter of 30,000 civet cats carrying the virus (ScienceInsider, May 4, 2009).

When asked why there hasn't been a SARS outbreak since 2003, consider how Yi describes that achievement: "The ecosystem [italics mine] was disrupted." Yi firmly believes that wild animals and birds constituted the SARS ecosystem, and therefore that slaughtering those creatures deprived the SARS virus of the environment it needed to survive. Perhaps we should follow Yi's eco-terminology and phrase the case for a Kyoto-style human health agreement in similar terms. The essence of the argument for legally mandating curbs on greenhouse gas emissions to arrest climate change is that carbon is a pollutant. When too much of this element is released by combustion, respiration, and other methods into the atmosphere, the result is global warming that is detrimental to human and animal welfare. Fair enough. It is only sensible to attempt to control carbon emissions. But why, then, have other carbon compounds, clearly more pernicious to human health and welfare, been left largely unregulated under international law? I refer here to viruses, malignant proteins, which are all carbon-based. Wouldn't it also seem sensible to control protein carbon emissions as well? Moreover, shouldn't the relative efforts made to combat nonprotein carbon and protein carbon emissions be in rough proportion to the amount of injury and death that leaving them insufficiently controlled might cause?

Control of infectious disease, particularly influenza, can and must be understood as an environmental issue. The flu hops from birds to pigs to humans to cats and back, changing and often strengthening with every leap it makes. Influenza is an ecological phenomenon. So why no Kyoto-style international convention regarding infectious disease and human health? WHO operates under the auspices of the United Nations and has very little legal power to enforce its guidelines and protocols.

There are no binding international laws governing the incubation or transmittal of viruses from nation to nation, no governing authority, virtually no legal or economic sanctions that may be imposed. Nonetheless, there is cause for hope. IHR (International Health Regulations) 2005, the long overdue overhaul of the original, obsolete IHR crafted much earlier, has given WHO a somewhat stronger hand.

"IHR 2005 ... constitutes one of the most radical and far-reaching changes to international law in public health since the beginning of international health cooperation in the mid-nineteenth century," writes Dr. David P. Fidler, a public health specialist at the Indiana University School of Law who has for years campaigned to strengthen WHO's legal authority. WHO's ability to declare an international pandemic and to recommend the imposition of travel and trade restrictions, though not strictly legally binding on member states, carries with it significant "bully pulpit" authority.

The bad news is that the expansion of WHO's powers by IHR 2005 in no way keeps pace with the expansion of the ability of viruses, influenza, and otherwise to propagate swiftly around the world. There needs to be an economic component, both incentives and penalties. Fidler has suggested that the World Trade Organization be allowed to impose trade sanctions against states noncompliant with WHO's infectious disease prevention guidelines. In times of infectious disease emergency, as declared by WHO, screening of airline and cruise ship passengers should become mandatory, just as we are physically and electronically screened for weapons and contraband. If your government is lax about letting people with flulike symptoms get on airplanes, then the rest of the world will tariff your ass off until you shape up, is the basic idea. Ditto if poultry in your live markets is harboring dangerous viruses and your nation's authorities refuse to slaughter and sanitarily dispose of these potentially infectious creatures.

AFTERMATH

The economic incentives to comply will be implicit but significant. Just as the Los Angeles County Department of Public Health posts big letter grades, A through C, right in the window of every eating establishment, so too should WHO be empowered to affix letter grades to nations and territories in times of health emergencies such as the latest epidemic of swine flu. It's a quick and easy way for travelers to determine where they should go for their business and pleasure, and where they should not.

Ultimately, the Kyoto Accord will be judged not only on its success in reducing GHG emissions and combating climate change, but also for how well it serves as a template for other challenging problems, such as the spread of infectious disease, that now more than ever need to be addressed by similarly comprehensive global legal and economic frameworks. It would also be nice if Kyoto does not, as naysayers vehemently contend it will, wreck the global economy.

CLIMATE CHANGE NAYSAYERS

Climate change naysayers are a different kettle of fish than us doomsayers. For the most part, they are cheerleaders whose basic belief is "the more the better." Their champion is S. Fred Singer, eighty-four, professor emeritus of environmental science at the University of Virginia and author of *Unstoppable Global Warming: Every 1,500 Years*, a *New York Times* bestseller stressing that climate change is an inevitable and benign force of nature. (Fred Singer is not to be confused with Fred Sanger, a British chemist who is only the fourth person to win two Nobel prizes.) According to Singer, man-made emissions of greenhouse gases are neither a threat nor readily controllable, nor should they be:

We're being asked to buy an insurance policy against a risk that is very small, if at all, and pay a very heavy premium. We're being asked to reduce energy use, not just by a few percent, but, according to the Kyoto Protocol, by about 35% within ten years. This means giving up one-third of all energy use, using one-third less electricity, throwing out one-third of all cars perhaps. It would be a huge dislocation of our economy, and it would hit people very hard, particularly people who can least afford it. For what? All the Kyoto Protocol would do is to slightly reduce the current rate of increase of carbon dioxide... by the year 2050... it would reduce the calculated temperature increase by 0.05 degrees C.

Singer is a gifted attack dog who takes singular delight in pointing out fallacies and errors in the climate change activists' work. Charging that the fact of academic life is that more grant money goes to those who adhere to the prevailing orthodoxy, Singer notes that one common practice is for scientists to faithfully report their findings in the body of a final document, but to slant the executive summary (which, after all, is the only part that most decision makers read) toward the political agenda of those providing the funding. For example, the executive summary of the landmark IPCC (Intergovernmental Panel on Climate Change) study on global warming issued in 1998 omitted the fact that weather satellite observations of the preceding twenty years had failed to indicate any warming whatsoever. Awarded a special commendation from President Dwight Eisenhower for his work in developing earth observation satellites, and subsequently made the first director of the National Weather Bureau's Satellite Service Center, Singer noticed that satellite data missing from the IPCC report summary were in fact buried in the main body of the 600-page report, which for some reason had no index. After Singer pointed out the discrepancy, the satellite temperature measurements were upwardly revised by the IPCC to indicate an increase in the world's temperatures.

AFTERMATH

Ironically, as a reviewer of the IPCC reports on global warming, Singer is a nominal corecipient of the 2007 Nobel Peace Prize shared by the IPCC and Al Gore, with whom Singer vociferously disagrees. Perhaps in hopes of countering Gore's 2006 film, *An Inconvenient Truth*, Singer participated in the controversial polemic *The Great Global Warming Swindle*, produced by United Kingdom TV4 in 2007. Charges of inaccuracies and intentional distortions have dogged that documentary, as has a salubrious notoriety.

Singer is a sharp-elbowed infighter who enjoys a good scuffle. Once he defeated Carl Sagan in a debate over the atmospheric impact of the oil well fires set off in Kuwait by Saddam Hussein's invasion. Sagan warned of profound danger from the blacking out of the sun, including massive agricultural failure and famine. Singer dismissed it all as a problem that would resolve itself in a matter of weeks; his prediction subsequently proved to be far more accurate than Sagan's. A contrarian at heart, Singer has doggedly disputed everything from the health risks of passive smoking to the link between CFCs and depletion of the ozone layer. His lot lies with big business; in 1964 he received a Gold Medal Commendation from the U.S. Department of Commerce, illustrating his lifelong penchant for allying himself with large and controversial industries, including oil, tobacco, and chemical interests.

Singer's colorful character and pugnacious stance on climate change may have put off some of his erstwhile ideological allies. Or maybe it was the climate change environmentalists' unrelenting barrage of "junk science" disparagement leveled at Singer and his ilk. Whatever the reason, it has only been recently, really since the 2007 airing of *The Great Global Warming Swindle*, that the ranks of the naysayers have begun to swell. Three scientists known as the Oregon Group published what has essentially become the science manifesto of those who challenge the importance of human-induced climate change.

"A review of the research literature concerning the environmental consequences of increased levels of atmospheric carbon dioxide leads to the conclusion that increases during the 20th and early 21st centuries have produced no deleterious effects upon Earth's weather and climate. Increased carbon dioxide has, however, markedly increased plant growth. Predictions of harmful climatic effects due to future increases in hydrocarbon use and minor greenhouse gases like CO_2 do not conform to current experimental knowledge," write Arthur Robinson, Ph.D., Noah E. Robinson, Ph.D., and Willie Soon, Ph.D., all of the Oregon Institute of Science and Medicine, in "Environmental Effects of Increased Atmospheric Carbon Dioxide," an extensive scholarly treatise published in the *Journal of American Physicians and Surgeons*, Fall 2007.

The Oregon Group firmly contends that atmospheric temperature is regulated by the sun, which fluctuates in output, and by the greenhouse effect, largely caused by atmospheric water vapor, which dwarfs carbon dioxide in importance. Systematically challenging every major aspect of the hypothesis that global warming is largely caused by hydrocarbon emissions, the group's survey of historical data found no increase in the number of Atlantic hurricanes that make landfall, nor in the maximum wind speed, nor in the number of violent hurricanes or tornadoes. Arctic surface temperatures correlate with solar irradiance, not hydrocarbon use. Sea levels have increased at the same rate regardless of the level of hydrocarbon emissions.

"Claims of an epidemic of insect-borne diseases, extensive species extinction, catastrophic flooding of Pacific islands, ocean acidification, increased numbers and severities of hurricanes and tornadoes, and increased human heat deaths from the 0.5 degree C per century temperature rise are not consistent with actual observations. The 'human-caused global warming'—often called the 'global warming'—hypothesis depends entirely upon computer model–generated scenarios of the future. There are no empirical records that verify either these models or their flawed predictions," charge the Oregon naysayers.

One eye-popping defection to the naysayer camp is Patrick Moore, a cofounder of Greenpeace, who was aboard the *Rainbow Warrior* on July 10, 1985, when it was bombed and sunk while protesting French nuclear testing at the Murora Atoll in the South Pacific. Moore has remained an ardent environmentalist, though he has since left Greenpeace, in part because of differences regarding climate change policy. He has come to the opinion that it is difficult to prove and even harder to arrest, and in any case not worth the cost, most of which will be borne by the poor of the world.

"I think one of the most pernicious aspects of the modern environmental movement is the romanticization of peasant life. And the idea that industrial societies are the destroyers of the world. The environmental movement has evolved into the strongest force there is for preventing development in developing countries. I think it's legitimate for me to call them 'anti-human,'" said Moore in *The Great Global Warming Swindle* documentary film. Moore's throwing in with the climate change naysayers gives them a potential hero figure, someone who put his butt on the line for an important environmental cause and who therefore cannot be dismissed or maligned without consequence, as wheeler-dealer Singer has so often been.

For intellectual stature, the naysayers now boast a towering polymath, Freeman Dyson, eighty-five, the mathematical genius dubbed "The Civil Heretic" on the cover of *The New York Times Magazine* (March 25, 2009). Dyson is best known academically for his contributions to quantum electrodynamics, which mathematically quantifies how matter interacts with light—for example, the way a sheet of transparent glass partially reflects, partially absorbs, and partially leaves undisturbed a ray of light. Although he does not hold a Ph.D., he has received twenty-one honorary doctorates from universities, including Oxford, Princeton, and Georgetown, has been a member of the physics faculty at Cornell, and now works at the Institute for Advanced Study at Princeton. In 2000, he won the $1 million Templeton Prize for science and religion.

Dyson believes that global warming is naturally cyclical, largely un-affected by human activity, and mostly benign. Like his allies, he attacks the climate change theory on the grounds of the "enormous gaps in our knowledge, the sparseness of our observations and the superficiality of our theories." Why pour so many resources into the as-yet-unproven problem of climate change, he asks, when there are so many undeniable problems begging to be addressed? A self-described humanist, Dyson contends that "protecting the existing biosphere is not as important as fighting more repugnant evils like war, poverty and unemployment."

Dyson is an unabashed fan of coal, which he considers cheap enough for most of the world to use, powerful enough to lift the masses of China and India from poverty to middle-class prosperity, and scrubbable enough to pass muster ecologically. His critics charge that no matter how smart Dyson may be, he has not done his homework on climate change. James Hansen, head of NASA's Goddard Space Flight Center and a standard-bearing environmentalist, sees the carbon dioxide issuing from coal smoke as the "dark agent of the looming environmental apocalypse," as *New York Times* writer Nicholas Dawidoff puts it. By contrast, Dyson rather likes CO_2. In his research monograph "Can We Control Carbon Dioxide in the Atmosphere?" Dyson speculates that the current rise in CO_2 emissions might be well worth whatever climate perturbations we are expe-riencing, because carbon dioxide helps plants, trees, and crops grow. In case CO_2 levels jump too high, he suggests that we could genetically engineer a special breed of trees that eat abnormally large amounts of carbon; planting about a trillion of them should tide us over until solar power technology comes on line in about fifty years. One wishes that rather than dreaming up fanciful tree-planting schemes, Dyson would turn his attention to solar technology. His original, groundbreaking work in quantum electrodynamics might now be updated and applied to teach us how light, in this case sunlight, can interact with matter, such as an advanced solar panel, to produce the greatest amount of energy.

The latest heavyweight to throw in with the naysayers is chemist Sidney Benson, ninety, a member of the National Academy of Sciences who earned his doctorate from Harvard University.

"Global warming is tremendously overrated. There is no evidence that it is happening, other than a slight, cyclical warming of the Earth. I would be very, very surprised if CO_2 correlates with global temperature," says Benson as we look around his small home office in the Brentwood section of Los Angeles. Winner of the Kapitsa Award, the highest scientific honor bestowed in Russia, and several times nominated for the Nobel Prize in Chemistry, Benson adds stature to the small but persistent group of global warming naysayers.

"The greatest threat we face from global warming comes from believing that it is caused by human activities. Poverty and misery will result from moving the global economy away from the fossil fuels such as oil and coal that have brought so much prosperity to so many around the world," declares Benson. He proudly informs me that on his ninetieth birthday he drove himself and his wife all the way across town to USC, where Benson taught for decades, to have lunch with Stephen B. Sample, the university president, and with his former student and lifelong friend, Ray R. Irani, chief executive of the Occidental Petroleum Corporation, who, I've since learned, once earned $460 million in a single year.

Sure, Benson's oil industry association casts doubt on his objectivity, although he did cheerfully volunteer the connection that I otherwise might have missed. But there's nothing inherently wrong with associating with the folks who provide the gasoline and heating oil on which damn near all of us depend every day. Nor does it necessarily mean that their facts are incorrect. What if Benson and, along with him, Singer, Moore, and Dyson are right, that cutting back on the hydrocarbons on which so much of our industrial and postindustrial civilization depends might needlessly plunge us back into an economic depression far deeper and more brutal than began in 2008?

If there's one thing that's been hammered into our heads since the OPEC oil embargo of 1973, it's that we have to save energy. It's expensive, it pollutes the air and water, it fosters our dependence on the Middle East, where many of our enemies are. It causes global warming. Yes, cutting back on our use of fossil fuels will diminish many of these problems, and yes, alternative energy sources such as solar and nuclear will likely flourish, but the question we never seem to explore fully (perhaps because deep down we don't believe it will ever really happen) is, what are the downsides of a shift away from the hydrocarbon-based economy? Disempowerment of the Middle Eastern oil states and their Exxon-Mobilesque partners? We can live with that. How about a jarring transition, perhaps lasting a generation or more, from an old, convenient way of life to a new and experimental one? Okay, but we're allowed to bitch about it.

What if things go terribly wrong and the powers-that-be fight back? Russia, the second-largest oil producer on the planet, would certainly want to stay the course, as, for that matter, might Venezuela, Texas, and Alaska. Given the precarious state of world affairs, with nuclear, biological, and chemical weapons spreading pandemically, we cannot discount the possibility of another world war. Are all the environmental and social benefits that will come from building an alternative energy economy worth suffering through another world war? Maybe.

Benson's warning about the perils of moving away from a hydrocarbon-based economy is oddly echoed by his ideological opposite,

AFTERMATH

James Lovelock, eighty-eight, a British atmospheric chemist whose work has shaped my career. Lovelock, who has been called the "prophet of climate change," believes that global warming is now past the point of no return, and that it will reduce the global population by billions of horrible deaths by the end of this century. Although he strongly advocates moving away from fossil fuels and toward nuclear energy, this changeover will not come without its own environmental cost. Lovelock observes that burning fossil fuels produces soot, which absorbs sunlight, thus rendering the Earth's surface cooler. Reducing soot would therefore raise surface temperatures, accelerating global warming by 2 to 3 degrees over the coming century. Anyone who has ever spent a sunny afternoon in the desert gets the idea.

How utterly confusing: two world-renowned chemists, fine men both, roughly the same age, same build, same gracious demeanor, both with lifetimes of superior work but with diametrically opposed worldviews, yet both warning us not to obey the First Green Commandment: Thou shalt not use fossil fuels.

Aftermath Scenario: Saved by a Banana

New Year's Eve 2012. The date shines like the full Moon in the reflected glory of the fabled Mayan end-date, 12/21/12, when the center of the Milky Way galaxy will be eclipsed by the Sun. To some, 12/31/12 flickers uncertainly, like the Moon passing in and out of the clouds. Will it be the first New Year's of a new era? Will it arrive at all? Others give the matter no thought, just as they take for granted that the heavens will always be there. But then there are those who focus on 12/31/12 as a different sort of deadline, the day when the Kyoto Accord is set to expire.

What had been a laborious exercise in science and diplomacy, two decades of meetings, signings, missteps, convoluted regulations, and jargon-laden communiqués, has now exploded in the popular imagination. For the Greens, Kyoto has become a watchword for our last, best hope of getting out of this mess. For the Carbons, as climate change naysayers have been dubbed because of their passion for oil, coal, and other hydrocarbon fuels, 12/31/12 could be the end of tyranny.

New and ambitious goals have been set for Kyoto II, the phase of the accord set to begin on January 1, 2013, but the Carbons do everything they can to delay the ratification process. The United States is markedly more cooperative than it has been in the past, but resistance from Russia, Venezuela, and much of the Middle East is staunch. Most of Asia, including China, India, and Japan, sit on the fence: on the one hand, they have little petroleum reserves of their own to defend, but they are disinclined to deviate from what has been their route thus far to greater prosperity and influence. Compliance with Kyoto II, even among signatory nations, is very difficult to enforce, because the sickly global economy seems headed for relapse; Wall Street warns that Kyoto II could knock it into a coma.

Why in God's name, the Carbons demand, should we choose this critical juncture to turn our backs on the greatest source of wealth the world has ever known?

For the Greens, economic concerns are trumped by fears for basic survival. The years 2011 and 2012 have proved exceptionally stormy and tumultuous, with natural disasters begetting political ones, governments toppling because of their inability to cope. Public opinion finds its scapegoat, blaming global warming for damn near everything that goes wrong in the world.

"Would you rather be poor or dead?"

That blunt question organizes the Greens. With surprising discipline and efficiency, a multinational cavalcade of politicians, business leaders, journalists, pop stars, and their supporters come together to purvey a single message: that it is time to slash fossil fuel use once and for all. Arguing that a few years of discomfort is preferable to unstoppable global warming, the Greens present a simple, radical agenda: 50 percent reduction in fossil fuel consumption over the next ten years. It's that, they say, or megadeath.

Conventional wisdom has it that negotiation is a process by which two or more extremes meet somewhere in the middle. But for both the Greens and the Carbons, the battle over Kyoto has become a crusade, and takes on a theological tone. Did God give us oil and coal to use for our benefit, or did He not? How about Allah? Not wanting to be branded atheistic, the Greens adopt a "pro-life" rhetoric, though in reality the movement is far more science-based than religious. "Thou shalt not kill," the sixth commandment in most liturgies, becomes the Green motto. This tactic infuriates antiabortion activists, who by and large toe the Carbon line.

Charges of corruption and influence-peddling leveled at the oil/gas/coal conglomerates backing the Carbons fail to stir much ire; the public just doesn't care. But neither does the man on the street believe the Carbons' claims of tremendous advances in clean coal technology, whereby the carbon-laden impurities that used to be scrubbed out of the air and then dumped into the water supply can now be sequestered and destroyed by plasma furnaces. Oddly, this ambivalence toward the Carbons is accompanied by fairly high approval ratings. The Carbons are uninspiring compared to the Greens, and, perversely, this works to the oil/gas/coal lobby's advantage. It's as though the public has hit emotional overload—whichever side invades their consciousness least gets their support.

What resurrects the Greens is a series of cataclysms throughout 2012, earthquakes, volcanic eruptions, and freak storms that shake the public awake. Even to the nonbeliever, it all just seems as though sent by angry gods, the same gods who sent the solar thunderbolts that disrupted the electrical power grid and satellite telecommunications. Climate change strikes with a vengeance as the surface temperature of the Siberian tundra crosses a threshold that triggers the release of approximately 400 billion tons of methane that for eons had been frozen inside underground ice structures known as clathrates. Central Asia is wracked with chaotic weather events, including hurricanes and tornadoes of unimagined ferocity, as well as accompanying seismic events. Agriculture, shipping, law enforcement, and military security are crippled in the region and, soon enough, around the world.

News sources reveal that a group of Russian scientists working in Novosibirsk, the capital city of Siberia, had long been predicting this outburst, strenuously arguing that methane monitoring programs are grievously underfunded and ill-conceived, leaving us ignorant of how much of this gas, with twenty times the greenhouse warming power of carbon dioxide, is being emitted around the world, or of what degree of warming might lead to a sudden, treacherous burst. The Russians had taken their research cue from a little-noticed finding in the groundbreaking 2008 United States Science Program study on ACC: "The size of the hydrate [methane and related gases] reservoir is uncertain, perhaps by up to a factor of 10. Because the size of the reservoir is directly related to the perceived risks, it is difficult to make certain judgment about those risks."

The Russians' work, like that of their American counterparts (whose study was funded, conducted, and published without comment by the legendarily pro-oil Bush-Cheney administration), had been ignored, passively suppressed by pro-Carbon regimes. The Carbons respond with copious evidence that it is not greenhouse gas emissions that are causing the climate change but rather the Sun, which climaxes with unprecedented power over the course of 2012. A few Green scientists acknowledge the Sun's growing role in climate change, but most of them take one for the team and ridicule the Carbons for trying to blame Earth's problems on Old Sol, our faithful energy provider 93 million miles away. The battle between the two camps sputters endlessly; no one wins and everyone loses because of it. A decisive victory by either side, particularly the Greens, would have been better than the directionless stalemate that muddled the second and third decades of the twenty-first century.

The only bright spot, in point of fact, a bright yellow spot, is the image of a perfect yellow banana that in 2012 begins popping up in videos, photos, drawings, and posters around the world. No credit is taken, no explanation given. Few pay the banana campaign much mind, and those that do figure it has something to do with a lingering dispute over genetically modified bananas: the demand for untinkered-with bananas has become too great and ecologically burdensome, but consumers, particularly Europeans, and the commissioners who represent them, refuse to accept the genetically engineered variety. The mystery banana catches on for a while as a pop fad, and there's even a slight uptick in banana consumption.

By the end of the year, however, the deeper intent behind this quirky campaign is revealed. The banana proves to be a symbol for an emerging new technology, much the way the apple made that brand of personal computers seem friendly and accessible. But what kind of technology does the banana represent? The second wave of banana images includes some information, that bananas contain potassium, a naturally radioactive element. And that bananas are good for you, not in spite of, but partly because of, this sprightly little fact of nature.

The banana turns out to be image advertising for a new generation of miniature nuclear fission reactors, being developed by companies such as Toshiba Corporation of Japan and Hyperion Power Generation of Santa Fe, New Mexico, which works in conjunction with Los Alamos National Laboratories. Depending on the design, these reactors will range in size from a hot tub to a mobile home and will generate power for up to forty thousand American homes or the equivalent. Buried an average of thirty feet underground, the minireactors will operate from five to twenty years, automatically, except, presumably, for the security squads patrolling the surface to defend against theft and terrorist attack.

Beyond the miniaturization, the key innovation in minireactor design is the use of a liquid lithium core, rather than control rods, to absorb neutron radiation. Neutron flow regulates nuclear fission much as oxygen flow regulates conventional combustion, that is, fire. At the first sign of trouble, liquid lithium should flood the fission reaction chamber, soaking up the neutrons and thereby shutting down the power. Manufacturers contend that they can retrieve these devices and recycle them after they run out of fuel. Radioactive waste is still a dilemma, though advances in plasma technology, whereby the furnaces are mobile and can destroy highly toxic wastes right where they are produced, mitigates this problem somewhat. The claim that no weapons-grade materials are produced by the minireactors is quickly substantiated by regulatory authorities.

Introduced in 2008, by 2012 the minireactors will have made it less than halfway through the Nuclear Regulatory Agency's evaluation process. Assuming the devices are approved without unduly burdensome regulations attached, another decade, at least, will be required to build, procure suitable sites for, and then install the minireactors. The goal of the banana campaign, whose backers were never positively identified, was to hurry things along, before the squabbling between Greens and Carbons does us all in for good.

SECTION III
HEADED FOR THE HEREAFTER

Writing about the Apocalypse from Beverly Hills has its challenges, especially around Christmastime, when Santa Claus rides down Rodeo Drive dispensing luxury designer gift packages from the backseat of his Rolls Royce convertible. Prophecies of doom just don't compute here. The only way famine will ever hit Beverly Hills is in the form of a fad diet, that's how most locals think. And if by chance hard times have the temerity to come our way, surely there's some budding designer/entrepreneur graduate of Beverly Hills High who would create an "Apocachic.com" line of apparel and accessories, make a quick bill(ion), and refloat everyone's boat.

But divine retribution is another story. Although the concept is out of fashion these days, it is not so different from the pop-spiritual formulation "What goes around comes around"—the wheel of karma versus the tantrums of Yahweh. Both imply a sense of moral balance and cosmic redress for wrongful acts, though with different emphases on what is punishable. Exponents of divine retribution might see Hurricane Katrina as having purposely been visited upon the party-hearty, anything-goes city of New Orleans—the twenty-first-century Sodom and Gomorrah, crushed by an angry God. An odious theology, though who can deny that if the Big Easy hadn't taken it quite so easy when it came to caring for its infrastructure, a little more ant, a little less grasshopper, then the flooding might not have been nearly so bad.

Jesus lamented that the word of God was "sown among thorns" and choked "by the cares of the world and the lure of wealth." (Matthew 13:18–23). Will God, fate, or the IRS therefore punish us for having charged up a storm at the local pet boutique, where doggie cookies sell for $3 a pop? Even if you can afford it, is spending ninety bucks to get your eyebrows plucked a luxury or a misdemeanor? This chronic, base-level fear of divine retribution can flare up at odd moments, such as the release of *Beverly Hills Chihuahua*, a Disney family comedy about how a pathetically pampered pooch wearing a diamond necklace gets lost and has adventures in sunny Mexico. No one would accuse this light-hearted film of being political, even though most of the local street dogs who risk their lives for this lily-white bitch from BH don't get so much as a Liv-a-Snap in return.

Coming out of the silly movie with my daughter, who was seven at the time, I did a double-take at the shiny silvery necklace with a large ruby-red pendant she was wearing. It was obtained by redeeming approximately 375 award tickets (cash value one cent each) from Chuck E. Cheese, a ringing, blinking kid's fun emporium that specializes in pupil dilation and Styrofoam pizza. There is no Chuck E. Cheese within the city limits of Beverly Hills. There are, however, seven-year-olds who actually do wear necklaces from Tiffany and Harry Winston, just like the Chihuahua.

Is it wrong, morally wrong, for a child, or for that matter a dog, to wear, say, $50,000 in jewelry? Not according to current free-market thinking. We seem to have reached a gentleman's agreement in our society that no amount of wealth is too much, as long as it is gained legally. Tycoons therefore have the right to lavish riches upon their children and pets because it's their money and they can do what they damn well want with it, even if these splurges are in bad taste, or indeed harmful to the objects of their affection. Besides, without the superrich, high-class jewelry makers and super-premium doggie cookie bakers would lose their jobs, and we would all be the poorer for it. No one prominent in our social debate argues that becoming superrich is just plain wrong, for the very good reason that no one wants to be called a hypocrite, since just about all of us, yours truly included, would roll in the dough if we could. The success of the superrich enables the rest of us to keep alive our dreams of the day when we could (but would never) spoil our own children, or our children's children's children, with trinkets from Cartier, not Chuck E. Cheese.

As any child, or dog, instinctively knows, the pressing question is not about right or wrong but about punishment. Who will get punished by whom for doing what, and how bad will it hurt? Will the superrich be scourged by a wrathful Almighty, or by greedy revolutionaries wearing "Eat the Rich" T-shirts? Is it time for Madame Defarge, the infamous revolutionary in Dickens's A Tale of Two Cities, to transubstantiate from literary character to flesh-and-blood operator of the (new, high-tech, complete with digital head-counter) guillotine? As with scores of other ultrawealthy communities around the world, Beverly Hills is not haunted by guilt but by fear, fear that the natural order of things must inevitably correct gross imbalances in wealth and well-being. The fact that homeless people sometimes get doused with lighter fluid and burned to death by stoned-out gang-bangers, as happened the other day to a harmless old alcoholic who lived on a street corner not far from Beverly Hills (the lowly crime didn't make much news) is not the billionaires' fault, nor is it ours. But the gap between those who are cozy in bed with their Frette linens and those who sleep next to fire hydrants just seems too wide.

AFTERMATH

Common sense, more than moral sense, tells us that if that gap doesn't close up some, we might all fall in.

Some have opined that if global catastrophe strikes in 2012, we will be getting what we deserve: divine retribution. For the record, I reject that mind-set as repulsive, though not necessarily inaccurate. Whether such retribution might in fact be the work of a deity, or simply the natural, homeostatic adjustment of a system out of balance, is beyond the scope of this book to determine. But no matter how the process works, it does seem that the 2008 stock market crash, credit crunch, and ensuing recession were just deserts for too much greed throughout American society and beyond, with the greatest culpability attaching directly to the greediest among us, on whatever rung of the socioeconomic ladder.

That money has replaced God in people's hearts is nowhere more striking than in Beverly Hills, where it is not, to the best of my knowledge, possible to purchase a Bible, a Book of Mormon, or a Quran. The only bookstore in the sister city of Cannes specializes in trendy, arty picture books, a number of which are the functional equivalent of soft-core porn. Yet my shiny little city is profoundly humane, with nice parks and lots of special events, a full schedule of programs for children and seniors, a fine library open every day of the week, even a white-glove recycling program wherein residents need not trouble themselves to separate their trash into different containers, because BH municipal employees and contractors perform that untidy task for us at the dump. The police are fantastic; average response time is about a minute. Once the wind blew open a door in my courtyard, setting off the alarm. When I got home, there was a nice note from the policeman who had checked out the problem. He stopped by next day just to make sure everything was okay.

Beverly Hills would be very high on the hit-list of religious fundamentalists who see our society, indeed most of the Western world, as irretrievably decadent, condemned to divine retribution for committing the sins of legalizing abortion, condoning gay marriage, banning prayer from public schools, and putting fluoride in the water supply. Muslim, Christian, and Jewish extremists do not subscribe specifically to 2012 or any other end-times date, although the nearness of the Mayan deadline does seem to resonate with their hope that the Armageddon/Apocalypse final battle of Good versus Evil is nigh.

What to do about these zealots? Thankfully, they are small in number, but like drug freaks on PCP, their strength seems superhuman at times. Attacking them frontally would only give them the battle they've been spoiling for. Ignoring them would risk overlooking the threat to global stability that they genuinely represent. Warning others that Armageddonists, Jewish, Christian, and Muslim, just might make their move to fulfill their sacred/profane mandate as the world builds toward the perilous climax of 2012 is helpful, as long as the warnings do not create runaway fear. Fear is warranted, even healthy, when danger is imminent. But sometimes fear blows out of proportion, like when the body's immune system overreacts to an infection, causing more harm than the infection ever would have alone. Although there is no greater threat to human safety than the all-out war to end all wars that these maniacs desire, runaway fear over the possibility that this might happen is not an effective response. Well-founded or not, fear can become its own form of danger, paralyzing those who would otherwise have been functional, or, worse, inciting panic and mayhem, thereby actually precipitating the danger that spurred it. The last thing we want to do is whip up hysteria about Armageddon and make it a self-fulfilling prophecy.

What to do about these zealots, shake our heads and shrug? Plato taught us that the way to defeat an opponent is to attack his argument's strength, not its weakness. The source of the zealots' strength is Scripture, so undermining that strength would help undermine them. Thus, I resolved to go to Patmos, Greece—where John the Apostle wrote the book of Revelation, the sacred, seminal text of Apocalypse and Armageddon—with the goal of discrediting the whole sordid notion.

SPELUNKING THE APOCALYPSE

don't like caves. Huge caverns, okay, they can seem like underground cathedrals. But anything smaller feels like a rocky, funky womb to which I have no fetal desire to return. Plus, one never knows what might be skulking inside. So why fly eleven hours from Los Angeles to Athens, then connect via death-cab to Piraeus harbor for a twelve-hour ferry ride, during which the squeaky fan above my bunk bed showered soot nonstop, all for the purpose of sitting in a cave? Because that cave, on the remote Aegean island of Patmos, is where John, author of the gospel bearing his name, is widely believed to have written in AD 95 the book of Revelation, the most widely known and deeply feared apocalyptic vision in history.

Going to the Greek isles to research divinity is, on the face of it, absurd. No place is more devoted to this-worldly pleasure, to the godless, hedonistic "whore of Babylon" whom John railed against. Yachts, nudity, and hook-ups galore—that's the currency of the place these days. It would be the perfect spot for the Antichrist to emerge . . . Is that a bottle of ouzo he just cracked open, or is it the Seventh Seal?

Dawn broke as the Blue Star ferry put into Patmos, and I could not help but think of the New Testament story in which Jesus foretold that

AFTERMATH

Peter the disciple would deny him three times before the cock crowed with the first morning light. Despite this cowardice, Peter was the rock upon which Jesus built his Church. A built-in flaw.

I doubt Jesus ever considered John for the job. John was more the philosopher type, from "In the beginning was the Word," as he opens his gospel, all the way to the bitter end: "Happy is the man who reads, and happy those who listen to the words of this prophecy and heed what is written in it. For the hour of fulfillment is near" (Revelation 1:3).

John was an educator, doctor, and painter, the first one to paint Mary's portrait. But with all due respect, it has been nearly two thousand years since the great gospeleer warned that the End was coming soon. Ten billion or so of us have loved, fought, dreamed, and explored since then, and with a little luck, 2012-wise, we have at least that much time or more ahead of us. Although sometimes known as St. John the Divine, the author was, after all, only human. Maybe John opened his prophecy tale so dramatically in order to attract attention, kind of the way I have done by including the 2012 end-date in the title of this book. Or maybe he was so stunned by his phantasmagoric vision that he just jumped to the conclusion that anything so vivid and terrifying must be close at hand.

In a nutshell, Revelation gives us the "what" without the "when," while the Mayan prophecy concerning 2012 gives us the "when" without the "what." Revelation provides a meticulous depiction of the events of the Apocalypse but only the vaguest clues as to when it all might occur. The Mayan prophecy is just the opposite, offering few details about what will happen as we emerge into a new era, while being quite precise about the date of 12/21/12 as the birth of a new age of enlightenment. Some will succeed in making the transition, others will fail. As it happens, the verse in Revelation that comes closest to capturing the make-or-break Mayan sentiment is, in fact, 20:12: "And I saw the dead, great and small, standing before the throne, and the books were opened. Also, another book was open, the book of life. And the dead were judged according to their works, according to the books."

Contemplating 2012 periodically forces one into denial, denial that global catastrophe could be on its way, denial of John's or of any other hideous depictions of said catastrophe, denial that this glorious life of ours could ever end. The fact that the 12/21/12 end-date is Mayan, a culture with which few of us have any direct personal connection, makes the prophecy seem both less probable and less personal. It's also more difficult for us to confront. We can ethnocentrically dismiss the Maya as a failed and/or marginal culture, but we can't take their arguments apart piece by piece since there is no central text akin to the Bible, or anointed spokesperson to confront. Revelation, on the other hand, comes from the same Judeo-Christian tradition from which most of us Westerners descend (although, ironically, the Bible is largely the product of the Middle East and the Mayan prophecies are geographically western).

Even if one does not believe in the power of evil incantations, it seems wise not to repeat them out loud. What if by chance there is some hypnotic power to the sounds of the words? Superstitious, I admit it, but that's kind of how I feel about Revelation; read it, analyze it, but don't invoke it, or "give energy" to it, in New Age parlance. Every time John's macabre story is creatively perpetuated, by Dante in the *Inferno*, by the grotesques of Hieronymus Bosch, by countless horror flicks, the cautionary value of the gory parable is offset, perhaps more than offset, by our wide-eyed fascination with it all. Just as many of us turn out like our parents, even if we try not to, so is human consciousness parented by certain immortal, and immortally flawed, texts. John's end-game prophecy has thrived so long in the human imagination that we cannot help but be unconsciously drawn to follow the familiar road map, even if it leads to Hell.

AFTERMATH

Poking holes in John's apocalyptic vision also pokes holes in the very idea that any apocalypse could ever occur anywhere. Refuting Revelation, that grandiose nightmare eternally hanging over the human horizon, would therefore be a service to us all. Scholars have tried. Some challenge its authorship, arguing that the John who wrote the gospel in flawless Greek is not the same John who later wrote Revelation, which has its share of grammatical errors. However, that could be attributed to the fact that Prochoros, the scribe who transcribed Revelation, was from a remote island and had little formal education. Others charge that it's just one of many apocalyptic texts that derive from the Old Testament book of Daniel. But there Revelation sits unscathed, in its honored place as the concluding chapter of the Bible, and of humankind.

SCORPION PLAGUE

Close your eyes, flip through the pages of Revelation, plunk your finger down at random, and odds are you'll hit on something like this: "Then over the earth, out of the smoke, came locusts, and they were given the powers that earthly scorpions have. They were told to do no injury to the grass or to any plant or tree, but only to those men who had not received the seal of God on their foreheads. These they were allowed to torment for five months, with torment like a scorpion's sting; but they were not to kill them. During that time, these men will seek death, but they will not find it; they will long to die, but death will elude them" (Revelation 9:3–6).

The best way to protect oneself psychologically, as most any of the pilgrims in Patmos would have known right off the bat, is to pray. I believe that praying actually does connect one with God or other deities, though I also know that believing doesn't make it so. Prayer may simply be the process of directing one's purest thoughts and feelings to the best and deepest part of oneself, consulting the "god within," even if such an entity exists only metaphorically. Whatever the ultimate truth of the matter, praying is a good thing to do, and is a superbly effective, and cost-free, defense in time of crisis. So before reentering the cave, I paused to thank God for bringing me there and prayed for His protection against the nightmarish images that had issued from within.

Prayer can backfire, though. One of the pitfalls is feeling self-righteous about it afterward, proud of being oh-so-humble. In one of those fool's-gold, "aha" moments, it struck me that Revelation isn't scary, it's ridiculous. Listen to the rest of what John (if in fact he is the one who really wrote it) has to say about those killer locusts in 9:7–11: "In appearance the locusts were like horses equipped for battle. On their heads were what looked like golden crowns, their faces were like human faces and their hair like women's hair; they had teeth like lion's teeth, and wore breastplates like iron; the sound of their wings was like the noise of horses and chariots rushing into battle; they had tails like scorpions, with stings in them, and in their tails lay their power to plague mankind for five months."

Mutant locust-scorpions assailing humankind to the point where we beg for death? That must have been some pretty potent cave gas John was inhaling, but what's Jeffrey A. Lockwood's excuse? In Six-Legged Soldiers: Using Insects as Weapons of War, Lockwood traces the history of man's use of insects as instruments of warfare back to the days when cavemen would throw a beehive inside an enemy's cave, then ambush their enemies when they ran out screaming. Entomological warfare reached new levels in the twentieth century, when Japan's Unit 731 "waged full-scale entomological warfare against China in World War II—and were on the verge of launching similar attacks against U.S. troops and the American public. Between plague-infected fleas and cholera-coated flies, nearly half-a-million Chinese were killed." Examine flies and fleas up close under a microscope and John's bizarre insect imagery does not seem nearly so far-fetched.

Lockwood's survey of emerging twenty-first-century insect weaponry could be drawn straight from Revelation itself. Supersensitive assassin bugs, bloodsuckers that can smell out and attack human beings even in the thickest jungles, have been bred for military use since the Vietnam War of the 1960s. Robotic bees have been developed by the "biomimetics" program of DARPA (the Defense Advanced Researched Projects Agency, credited with having funded the basic technology of the Internet). Robo-bees, easily equipped with poison stingers, apparently now have the ability to attack both individually and as a swarm. And that's only what DARPA is telling us about.

John's prophecies of plagues and other natural disasters just don't seem so archaic or improbable after reading Lockwood's work. Sure, John most likely confused some, maybe most, of the details, but there is one very consistent theme in Revelation: *Nature will be turned against itself and therefore also against us.* Locusts attack human beings, seas boil with blood, frogs run amuck. What even fifty years ago might have seemed quaint flights of fancy that paled when compared with the real technological threat of atomic weapons, space wars, and such, today are a lot more chilling, as bioengineering, at its most perverse, proceeds to realize John's immortal nightmare.

For the first time since Revelation warned of the killer locust plague, such a scenario is possible without supernatural intervention. But that still doesn't mean we have to buy John's whole doomsday thing.

GETTING HIGH TO SEE GOD

I felt like Judas later that morning as I entered the Cave of the Apocalypse, as it is called in Patmos. The cave can hold around forty tourists at a time, but when I visited in early May, only a dozen or so poked in their heads over the course of several hours. I sat in a chair next to the holy crack in the wall from which emanated God's word to John, who, in turn, dictated the words to Prochoros, his faithful scribe. After twenty or thirty minutes of just sitting in the cave, taking it all in, I experienced my first miracle: that skinny, rickety, hard-assed little chair became as comfortable as a Barcalounger. I could have sat there for the rest of my life.

I think I was high on cave gas. Decreased oxygen levels and elevated carbon dioxide, nitrogen, radon, and hydrogen sulfide are conditions typical in caves. When carbon dioxide levels climb above 2 percent of the total available air supply, effects on respiration and pulse rate start to become noticeable. And when CO_2 reaches 3 percent of the available air supply, as it does in a number of caves, delusions begin to occur.

Maybe John got high on cave gas and hallucinated his head off for the two years he was living there. That would be a novel line of criticism, that Revelation was written by a guy stoned out of his mind. (Certainly would explain all those grammatical errors.) Then again, caves have incubated some of the most miraculous works of literature in history, most notably, the Quran, which was revealed by Allah via the Archangel Gabriel to the Prophet Muhammad during a series of meditations he conducted inside a Mecca cave. By Islamic tradition, known as "hadith," Muhammad was illiterate, although some scholars allow for the possibility that he could have acquired rudimentary reading and writing skills while traveling with trade caravans to Jerusalem. In any case, there is no ready explanation, none, other than divine guidance, for this simple tribesman producing the greatest masterwork in the history of the Arabic language.

If meditating in caves was good enough for Muhammad and John, then it was good enough for me. What reason could I possibly have for doing anything less than the same to discover the truth about 2012?

AFTERMATH

Sheer terror, for one thing. Terror of exposing my psyche to perversion, suffering, mania, and death. Revelation is full of seas boiling with blood, the earth being burnt to a crisp, water turning to poison, horses with lions' heads and tails like snakes, squadrons of killer angels, unrelenting agony. Regardless of whether Revelation is truly anything more than just a random, hellish diatribe, the stain of its imagery has been setting in our collective unconscious for two thousand years. If for no other reason than just the power of suggestion, meditating in the cave where all this sick stuff started seemed a psychologically risky thing to do, so I struggled out of my rickety, comfy chair and stepped out for some fresh air.

Unlike Daniel Pinchbeck, a connoisseur of hallucinogens whose commentary on 2012 I find beguiling, any writing that I've ever done under the influence of intoxicants could be charitably described as somewhere between purple and drivel. However, the one and only time in my life I took LSD, I found the word of God, literally. Wandered into a bookstore and discovered the Gospel According to Thomas, thought of by some as the "fifth gospel," essentially the sayings of Jesus, culled from an ancient papyrus known as the Nag Hammadi, discovered in Egypt and certified by an international collaboration of biblical and linguistic scholars. Up until the discovery of that ancient manuscript, the only speaking role for Thomas in the New Testament came in John 20:25, where "Doubting Thomas" insists on putting his fingers in the holes of Jesus's crucifixion wounds.

In *Beyond Belief: The Secret Gospel of Thomas*, Elaine Pagels contends that Thomas and his followers, known as the Gnostics, were also skeptical about the physicality of the Rapture, the end-times phenomenon when Jesus returns to Earth and the faithful accompany him to heaven, regardless of whether they are living or dead. The Gnostics believed that only our souls, not our bodies, will be raised when Jesus returns. John, however, explicitly foresaw the emptying of the graves.

Mayan prophecy for 2012 includes no flying corpses, I am happy to say.

According to Pagels, Thomas would not have read Revelation as a literal prediction of how the battle between Heaven and Hell will play out in the actual physical world. Rather, he would have seen it as an artistic, symbolic, impressionistic rendering of the struggle between Good and Evil that happens in our souls. If Revelation is nothing more than John's own (cave-gassed) psyche writ large for us all to relate to, then what's the worry? If correct, Thomas and his followers defuse the whole issue to the blessedly uncataclysmic psychological realm.

The Gnostics are the ideological opposites of the Orthodox, who have no use for rhetorical loopholes, such as that the word of God is largely metaphorical and therefore open to subjective interpretation. Besides, it wasn't Thomas's cave I reentered, it was John's. After an hour of praying and meditating in the cushy little chair, I saw in my mind's eye three sharp-toothed serpents pop out of the back of my head. Frightened, I opened my eyes and they disappeared. But when I closed my eyes again, the snakes were still there. Opened and closed yet again, and the snakes were still there. Not a hallucination exactly, though neither was it just a dream. The middle snake arched its head back and, like a Pez candy dispenser, out of its neck came a glowing pellet, the color of sunshine. Serpent Pez was tempting me. The risk was mine to run. Should I pluck out the candy and pop it into my (imaginary) mouth?

Of course, there was probably nothing divine or supernatural about what I was experiencing, just self-suggestion hyperstimulated by the long journey and evocative setting. Still, I had come seeking revelation, and got one of sorts: when I was offered the possibility of revelation, in the form of that piece of candied sunshine, I declined.

Okay, if there's one lesson we learned from the Bible, it is not to accept tasty gifts from snakes. Turning down Serpent Pez should have been a no-brainer. But please understand that we journalists are not just nosy, we're self-righteously so—it is our duty to poke our noses where they do not belong. Professionally speaking, it is unethical, or at least poor form, to turn down an opportunity to find out or experience something that might inform or entertain the reader, or oneself for that matter. Beyond avoiding the Devil, in whom I do not believe but nonetheless fear, there was a deeper reason for shunning what would doubtless have been a mesmerizing cavalcade of Revelation-related insights and images: it would have been wrong. I had come to this holy place intending to prove that the scripture that had emanated from it was false, so I was the serpent, I was the treacherous one. Prayer and cave gas had done their job, connecting me with either the god or the devil within.

THE BE-GOOD SCENARIO

Archimandrate Antipas, the abbott and patriarchal exarch of Patmos, received me the next day. Antipas took his name from Revelation 2:13, as God's faithful witness who was murdered in Pergamum, the city where Satan was said to have his throne. Antipas is the sovereign of the Holy Cave of the Apocalypse and also of the Holy Monastery of St. John the Theologian, a magnificent nine-hundred-year-old complex built several hundred meters up from the cave, on the highest point of Patmos island. Both cave and monastery received their certification as UNESCO World Heritage sites while under his authority.

Antipas knows the book of the Apocalypse, as he calls it, by heart in the original ancient Greek and has written learned commentaries about it. To get to Antipas, however, one first has to get past Elder Mark Prochomios, his lieutenant who, when informed that I had once written a book named after Gaia, the Greek goddess of the Earth, immediately adjudged me as New Age, and therefore Satanic, albeit unwittingly so. As there is nothing quite so attractive to a clergyman as a sinner in need of redemption, Prochomios proceeded to lecture me about the evils of paganism. An hour later I was led into an ancient, impressive office to have my audience with Antipas, in his high, flat-topped black hat, great grey-flecked beard, flowing black robes, and sparkling silver ankh medallion. Although I don't understand Greek, I'm sure I was introduced as the goddess worshipper from California.

"Plant more trees. Use less machines," said Antipas, nodding amiably and extending his hand to be kissed. Prochomios, the lieutenant, would have collapsed at the New Age heresy of it all, were it not against the rules.

When I asked Antipas where we are today in the unfolding of the Revelation chronology, he reminded me that the gospel says that no man shall know when the End is nigh, referring to the quotation of Jesus in Matthew 24:35–36: "Heaven and earth will pass away; my words will never pass away. But about that day and hour no one knows, not even the angels, not even the Son, only the Father."

I was ready for that one, and responded that while Jesus proclaimed that no man would know the exact date when Heaven and Earth would finally pass away, he made no such declaration concerning the events, such as those described in Revelation, that would lead up to the end of time. In fact, He cited certain signs, such as the Sun and the Moon darkening, the stars falling from the sky, people making lamentations, that would indicate that the end is near. So I repeated my question: where in the end-times process spelled out in Revelation does Antipas think we are? Antipas retorted that John's apocalyptic vision is not so much symbolic or metaphorical as it is highly encoded to conceal it from those unprepared to receive it, broadly implying that few people outside of the Greek Orthodox clergy, including me, are emotionally, psychologically, and spiritually prepared for the news. I retorted I had no desire to know that which I had no right knowing, and gave as evidence the story of declining the glowing piece of candy offered me by Serpent Pez. Admittedly, it was a gamble, confessing to the hoary clergymen that Satan and I had had an encounter in their holiest shrine.

Prochomios rose to boot me out, but Antipas spared me with a gesture, then asked if I take any prescription medications. I couldn't tell if his question was a reference to the serpent's candy, or a more general commentary on my apparent psychological state. Neither. Antipas had segued into the point that every time one agrees to take a prescription medication, particularly one that must be taken indefinitely, such as Prozac, the mood-lifter, or even cholesterol-reducing Lipitor, one surrenders some of one's freedom to the pharmaceutical company whose product has now become necessary to one's existence. Sometimes that's necessary, he acknowledged, but only as the very last resort.

Satan ensnares, entraps, and fosters dependencies, is how Antipas sees things. The conversation drifted to the Antichrist, and then on to the Mark of the Beast, from Revelation 13:14–18, an imprint on the right hand or the forehead, bearing either the name of Satan or his number (666), without which, according to John's nightmare scenario, one could not buy or sell anything. Microchip tracking devices of the kind now being implanted in pets, livestock, inventory, and an increasing number of workers, and now even some schoolchildren, were cited with a shudder by Antipas as evidence of John's beastly prophecy. (Oddly, the same dread of microchip implants came up in my interviews with shamans and elders from Siberia, Guatemala, and the Hopi reservation in Arizona.) Each RFID (radio frequency identification) device emits a unique radio frequency, and can therefore be used to monitor the wearer's location. Already worn as ID necklaces by schoolchildren in several districts in Japan, and also in Sutter, California, where school officials plan to integrate barcode capabilities that enable the wearer to purchase goods, most RFIDs are the size of a grain of rice or smaller. According to In-Stat, a technologically oriented market research firm, the number of RFIDs produced annually will have risen from 1 billion in 2006 to 33 billion in 2010, and will then fall to zero, I hope, before 2012.

Evil, progress, or both? In terms of Revelation, the Mark of the Beast reference comes at the end of the penultimate section, just before the conclusion, "Visions of the End," which contains the first and only mention of Armageddon. I asked Antipas if the emergence of these tracking chips indicated that we were therefore at the beginning of the end-times war of Good versus Evil. He declined to disagree with my assessment.

VISIONS OF THE END

If what I gathered from my meeting with Abbott Antipas is correct—that all the credit cards, Internet cookies, and spyware recording our interests and purchases, the GPS chips in our cell phones tracking our whereabouts, and now those RFID implants, together constitute the Mark of the Beast—then what's coming next is wretched beyond belief: the seven plagues, earthquakes demolishing great cities, monumental storms pulverizing people, the great Armageddon war of Good versus Evil, anguish, falsehood, betrayal, men gnawing their own tongues. Basically, every horrible thing that could possibly rain down upon us or bubble up from beneath us is set to do so, starting pretty much now.

What is God so angry about? Lack of allegiance to Jesus, of course, this being the New Testament, and sinful behavior, mostly lust and greed. Uh-oh. Lust has pretty much been removed from the immoral list these days; it's 24/7 on basic cable. And what's so wrong about Victoria's Secret, Viagra, or both? Does anyone who has had really good sex ever really regret it, much less repent? Murder, we all agree on, and theft, false witness, and honoring our parents. Cutting back on the graven images would be hard, since that includes Internet, television, and films, which tend to fall in the lust categories. Keeping the Sabbath holy, that is, not engaging in commerce, will be difficult, though it might be a good way to save some money. But the finale of Revelation doesn't open with any of those:

> Then I looked, and on Mount Zion stood the Lamb, and with him were a hundred and forty-four thousand who had his name and the name of his Father written on their foreheads. I heard a sound from heaven like the noise of rushing water and the deep roar of thunder; it was the sound of harpers playing on their harps. There before the throne, and the four living creatures and the elders, they were singing a new song. That song no one could learn except the hundred and forty-four thousand, who alone from the whole world had been ransomed. These are men who did not defile themselves with women, for they have kept themselves chaste, and they follow the Lamb wherever he goes. They have been ransomed as the firstfruits of humanity for God and Lamb. No lie was found in their lips; they are faultless. (Revelation 14:1–6)

One hundred forty-four thousand virgin choir guys of perfect moral character . . . even if we degenderize the text to include women, raise your hand if you qualify.

Greed has become more or less synonymous with good financial management, the last stock market crash notwithstanding. True, we have begun to move past the "Greed is good," Gordon Gecko days of Wall Street film fame, though that may only be because we've had our pockets picked by those greedier than we ever were. Still, straight up, would you trade a life of wealth and plentiful orgasms for godliness and virtue? The Beast, aka Satan, would be ecstatic at the results of that poll. There has never been a better time in history for Satan to brand us forever as his.

In John's foretelling, the whore of Babylon gets stripped, beaten, and burned to ashes: "The woman was clothed in purple and scarlet and bedizened with gold and jewels and pearls. In her hand she held a gold cup, full of obscenities and the foulness of her fornication; and written on her forehead was a name with a secret meaning: 'Babylon the great, the mother of whores and of every obscenity on earth.' The woman I saw was drunk with the blood of God's people and with the blood of those who had borne their testimony to Jesus." (Revelation 17:3–6)

What's so bad about being rich and lusty? Isn't making and spreading wealth, and along with it, physical pleasure, good for the soul, as well as the body? We live longer, look better, eat better, move stronger and faster, think harder, create more, and have better sex than ever before. For this very human culmination of the divine gift of life, God would destroy us? Methinks He might be insecure.

Poking holes in Revelation is like poking holes in Swiss cheese. What's the point? It doesn't really change the look, taste, or quality of what you consume. Sure, the last chapter of the Bible could be nothing more than a projection of John's personal psyche, perhaps heavily under the influence of cave gas. Or it could be a literal description of what might occur, and still be dead wrong. The moralizing does seem outmoded, as might well be the notion of God as an angry, jealous, all-powerful man in the sky. But after my pilgrimage to Patmos, I came away with a grudging affection for John, kind of the way one could not help but care for the television character Archie Bunker, not because of what he said or did but because he cared enough to yell and kick when he thought something was wrong. Maybe we do worship the dollar, euro, pound, and yen too cravenly, and maybe we should keep it in our pants a little more. If not, as Archie said and as John would agree, God just might "kicketh the crapola" out of us, and send us to Hell.

HELL

Civilization collapses, great truths are revealed, judgments are made, the faithful ascend to Heaven and the rest of us go to Hell. That's the upshot of Revelation. But who believes in Hell these days? The traditional concept of Hell as an actual physical place to which unrepentant sinners are consigned for eternity is laughably out of fashion. We deserve a break today, not eternal damnation. And as for Satan being an actual physical being, no one thinks that anymore. A mythological character of extreme poignancy, the ultimate metaphorical embodiment of evil, and possibly even, in some dimly understood metaphysical sense, an actual motive force for sinful conduct in this world—that's about as far as most folks will go.

It's easy to ridicule the ham-handed constructs of Revelation and other fire-and-brimstone biblical texts; good sport, actually. But nitpicking the details does not mean proving the case that Hell does not exist. What is Hell then, if not a physical place? One certainly cannot dismiss the grim possibility that Hell exists as the permanent loss of one's soul, eternal separation from God. It's a bone-chilling thought, lonely and void; Judas in the lowest circle of Dante's *Inferno*, stuck upside down forever in the ice.

A more productive approach to the whole concept is that the reasons for which one might be condemned to Hell keep changing over time. Recall Huckleberry Finn's heroic decision to help his friend Jim, a black slave, escape to go find his wife and children, who had been sold down the river. Huck believed he would go to Hell for doing that because back in the pre–Civil War era in which Mark Twain's story was set, slaves were considered property, thus making Huck an accomplice in violating the commandment "Thou shalt not steal." If, in God's judgment, Huck did the right thing, would a God-fearing person who felt sympathy for Jim but who refused to help him because it was a violation of the command-ment have therefore been damned?

After the Civil War, slavery was abolished and theology changed. In *The Diary of Henry W. Ravenel*, an obscure but compelling Civil War memoir of a South Carolina aristocrat, a self-styled gentleman botanist, Ravenel, found himself puzzled as to why God had allowed the South to be so badly destroyed. He acknowledged the evils of slavery, specifically the practice of selling family members to different owners such that they would never see each other again. But Ravenel, a good man, it seemed, firmly believed that, in time, the odious practice that Twain wrote about would have been phased out. He also knew that God knew that decency would prevail eventually, and was perplexed as to why the Almighty didn't show more patience before raining down apocalypse, in the form of the War of Northern Aggression and then Reconstruction, on his beloved Confederacy.

Although the white slave-owner's excuses today seem pathetic and re-pugnant, the purpose here is not to heap scorn in retrospect but rather to wonder what manner of institutionalized evil have we just taken too long to remedy, so long that apocalypse or some functional equivalent thereof just might drop on our heads. I asked Antipas about changing morality, citing the evolution of standards regarding a woman's right to disobey her husband and/or to pursue a career. Antipas responded simply that the Ten Commandments specify nothing regarding the dynamics of male-female relations other than a ban on adultery, which applies to both genders.

Goddess-worshipping Californian that I apparently am, I wondered aloud if it weren't time for a new commandment regarding "ecological sins," including, for example, the wanton despoiling of a precious rain forest. Any chance of that becoming a new line item on St. Peter's balance sheet? More urgently, is there any reference point in Revelation indicat-ing the wholesale destruction of the natural environment, as now is happening far more rapidly than ever before, as a stage in the downward spiral toward fulfilling that frightening biblical prophecy? Antipas encouraged me to take a more positive view, reminding me that Revelation is ultimately a joyous text. After the monster battle that destroys the great city of Babylon, along with much of the rest of the world, things settle down and Jesus takes over. This happy scenario is akin to the Mayan vision for the emerging Ethereal Age, which starts out rocky but then moves toward enlightenment. The Mayan timekeepers project five thousand or so years of higher, happier consciousness, as opposed to the Revelation version, in which Christ's reign ends after only one thousand years, and Satan is once again let loose to run amuck until he is vanquished forever after yet another cosmos-boggling battle.

After our conversation, Antipas graciously invited me to join him for lunch in the monastery, which was delicious, though silent, except for some prayers and readings, all in ancient Greek. Antipas is a placeholder. He, or someone who looks, thinks, and prays a lot like him, will be right there in that same monastery, eating the same fresh local seafood, reading and rereading the same ancient Greek texts, and saying the same Orthodox mass, until the day comes when civilization really does crumble for good.

"For my part, I give this warning to everyone who is listening to the words of prophecy in this book; should anyone add to them, God will add to him the plagues described in this book; should anyone take away from the words in this book of prophecy, God will take away from him his share in the tree of life and the Holy City, described in this book" (Revelation 22:18–19).

A TALE OF TWO HEMISPHERES

a Tale of Two Hemispheres? Is that what Dickens would have entitled his book about 2012? Instead of London and Paris, would the story have taken place in New York and Beijing? Perhaps his underlying message would have been that in this hyperglobalized era of ours, hemispheres don't count anymore, that the two halves of the planet are as intricately interdependent, as hopelessly integrated, as the two halves of the brain. That we're all in this together, united more deeply than ever before as a civilization, as a species, facing the common threat of apocalypse, and/or revelation of the divine. But Dickens just might have been wrong.

In my previous book, I reported that the Maya expect 2012 to be a profoundly unsettling year of unprecedented historical importance. I have since come to learn that this expectation is echoed in the prophecies of many other native cultures throughout the Americas, including but certainly not limited to the Hopi, the Mohawk, and a number of Peruvian and Bolivian indigenous tribes. All in all, a gloomy portent. When beginning research for this book, I latched on to the hope that indigenous belief in the significance of 2012 might not be global but only regional in scope, ultimately sourced from the same (potentially erroneous) ancient Mayan texts. I therefore decided to conduct a "2012 check" on prophecies from indigenous cultures in other parts of the world, specifically the Eastern Hemisphere. My reasoning was that if far-flung seers, prophets, and sages had indeed arrived independently at the same basic, cataclysmic conclusion, then it really would be time to get out the crash helmets . . .

EAST IS EAST

It was one of those trips where everything went wrong but turned out better than if everything had gone right. During my voyage from Los Angeles to Irkutsk, a Siberian city near the Mongolian border, in July 2008, every connection was missed, every lead turned cold, every contact dead or disappeared, every expense doubled—250 bucks a night for a hotel room on a muddy street in Podunksky, Siberia. But dumb luck, everyone's best friend, saved my bacon. Purely by chance, the Buryat Festival, which every other year gathers indigenous people of Mongolian heritage from all around the world to meet for a long weekend, was being held in Irkutsk for the very first time. Athletes, artists, intellectuals, and young folks all come to disco, and they all could have been from the Navajo or Inuit reservations or, for that matter, from any one of the Mayan towns in Yucatán or Central America. It is startling, actually astonishing, how much the Mongolian/Buryat people look and sound like Native Americans, square and sturdy, with broad, weather-beaten faces, high, flat cheekbones, deep, dark eyes, and a calmness that made some seem timeless, others, oblivious. The theory that the groups somehow connected by crossing over the Bering Sea land bridge twenty thousand years ago has to be correct. Perhaps the 2012 prophecy was also carried in their genes . . .

Normally, shamans don't attend such events. However, Irkutsk is the closest major city to Lake Baikal, the largest, deepest, oldest, and purest lake on the planet. Twenty percent of the world's fresh water supply, equivalent to one-fifth of all the water anyone drinks, cooks with, bathes in, or uses to water the lawn or the crops, is stored in that lake. Shamans from around central Asia therefore regard Baikal as holy, so dozens of them, of Mongolian and Siberian heritage both, turned out for the event.

"Shaman," it turns out, originally comes from the language of the Evenk, an indigenous people of north-central Siberia. Siberian shamans are widely regarded as especially powerful. But what do they do, exactly?

"Shamanism is a special form of religion representing a system of world ideology, immense in terms of its integrity and denoting all spectra of man-environment relations. As a special system of ideology it includes some elements of rational, irrational and artistically descriptive world perception," writes L. M. Kolyesnik in *Shaman's Costumes*, a monograph from Irkutsk Museum of Regional Studies, 2004.

Shamans act as intermediaries between humanity and nature. Their basic function is to keep us connected to the natural world and in so doing to heal us physically, emotionally, and spiritually through rituals that include meditation, singing, manipulation of earth, air, fire, and water, and occasionally the taking of natural medicines. Much of this is smoke and mirrors, but for the legitimate cause of breaking us out of harmful mind-frames by pushing us to physical extremes and in so doing restoring our sensory connection with our natural environment and ultimately with Mother Earth as a whole.

Most shamans are vitally connected to the land where they live, almost as though they are human emanations of their surroundings. This connection gives them an extraordinary sensitivity to the natural world, although their narrowness of locale calls into question their capacity for broader judgments, especially ones broad enough to reliably assess 2012. Nonetheless, it struck me that if shamans in Siberia, half a world away from Guatemala, were picking up the same basic "messages" from Mother Nature concerning 2012, that would be an important indication.

Specifically, I sought to know what these shamans thought about the Mayan prophecy for 2012. The good news is that those whom I interviewed do not regard 2012 as the end of the world. The bad news is that they do see it as a compelling avalanche date for Western civilization, basically the lands dominated by Muslims, Christians, Jews, and atheists of similar extraction, extending west from the Middle East to the Pacific Coast of North America.

"Do I believe in apocalypse? No, not for the whole world, though I think it may be time for the West to fall," said Nadezhda Bazarzhapovna, a professor of anthropology at the Academy of Culture in Ulan-Ude, a Siberian city close to the Mongolian border. Though not a shaman herself, she works as a liaison between the university and shamans throughout the region, and has therefore worked with a score of them over the past two decades. She sees contemporary Western civilization as being close to a bitter reckoning, if not in 2012 precisely, then thereabouts.

"Shamans connect us to our ancestors. Eastern people have clearer souls than westerners, because easterners have kept faith with their ancestors." Westerners, in her opinion, fail to honor those who gave them rise. Those of us who have kept faith with our ancestors will survive and transcend the turmoil of 2012, and those who have not done so will not.

That's one I didn't expect. What's the big deal about honoring ancestors, who are dead, after all? Is 2012 some sort of revenge from the grave? No doubt there are certain advantages to ancestor worship. If you grow up honoring your ancestors and seeing others do the same, you have confidence that, if nothing else, even if there is no afterlife, your memory will live in the hearts, or at least the rituals, of your descendants. The disadvantage may be that progress and innovation—any obvious break with the past, particularly in social mores—are not embraced as avidly as in the more forward-looking, ahistorical cultures of the West. Ancestor worship tends to slow social progress, sometimes to the point of bottleneck, until, that is, radical breaks are required, such as Mao Zedong's Cultural Revolution.

"East is East and West is West and never the twain shall meet." As Bazarzhapovna spoke, I couldn't help thinking of the quaint old Rudyard Kipling quotation, and of how outdated it seems these days. Likewise her warning about us not keeping in contact with those who are dead and buried. How, in this breakneck era of technological and cultural change, with new social paradigms sprouting every year, could it be life-or-death important to bow to the outmoded past?

"You have forgotten how to mourn," she explained. Mourning is a survival skill that will become ever more crucial with the many untimely deaths of family and friends that occur as catastrophes escalate toward 2012. Bazarzhapovna's opinion that the West will suffer more than the East was not so much a value judgment as a practical assessment of our combat-readiness. Cultures which practice ancestor worship, most notably, in China, will have the comforting advantage of ancient, familiar rituals through which such contacts may be established, grief released, and recovery made.

Asian shamans are generally quite dedicated to the principle of ancestor worship, one of their main functions being to help keep people connected to those ancestors, enabling adherents to maintain "clear souls." In a nutshell, ancestor worship emphasizes the influence upon the living of the dead, who are seen either as benevolent forerunners blazing a trail for their living descendants, or as vengeful, disturbed spirits who must be placated, or both. Ancestor worship is sometimes known as family worship, since its net effect is to increase awareness of and reverence for those who gave one rise. This spiritual tradition and mind-set has shaped much of Africa, Polynesia, indigenous American cultures, and, most notably, China, which by consensus is rising in global influence. It therefore behooves us to know more about ancestor worship and to consider how its basic tenet—respect for the dead—could apply constructively to our lives today.

One obvious relevance to 2012 is the superior nature of the mourning rituals found in cultures that engage in ancestor worship: elaborate burial systems, repeated visits to graves and monuments, rituals conducted both at home and at temples and other religious gathering places. Though performed as an act of respect for the deceased, the complex system of mourning practiced in ancestor-worshipping cultures serves the living as well, releasing grief, fear, and other negative emotions and reassuring the living that they will not be forgotten when their time comes. In conjunction with the more conventional, Western, psychological field of grief counseling, improving our knowledge and skills of ritualistic mourning will prove to have been a wise move should 2012 live up to its advance billing as a tumultuous year.

Several days later, I met a serenely powerful woman, call her Netsere, from outside Ulan Bator, capital of Mongolia. (For security reasons that will become clear, none of the shamans I met in Siberia can here be identified or described physically.) Through a phalanx of interpreters, English to Russian to Mongolian, I learned that Netsere agreed with the basic assessment that 2012 is not a date of particular significance to central Asians, though it struck her as about the right timing for the fall of the West to really begin to accelerate. Netsere emphatically agreed that lack of respect for our ancestors had put Western civilization on the road to perdition.

"Go deeper, look backward, be humble," advised Netsere.

"Fat chance," was my, and the rest of the Western world's, (silent) reply.

From the central Asian shamans' perspective, the year in history most analogous to 2012 would probably be AD 456, a date that culminated the centuries-long process of the fall of the Roman Empire and the beginning of the Dark Ages. They would almost certainly have seen the United States as the center of today's Roman Empire, and would likely have confirmed that we are headed for our own decline and fall. Will the capital that crumbles in 2012 be Washington, D.C., and will darkness once again descend upon the world? Or will relief from the burden of command spawn an American Renaissance similar to what happened, albeit almost a millennium later, in Italy, the nation-state descended from Imperial Rome? It is germane to recall that, in the context of Asian history, including China, India, Siberia, as well as much of the Pacific Rim, the fall of the Roman Empire was more than a footnote, though not much more.

Like other shamans I have met, Netsere was less interested in generalizations than in fulfilling the potential of the moment. She had an intensely personal focus, and soon she was probing me in ways that I, as the journalist and normally the asker of questions, was totally unprepared for.

Netsere saw me as someone with a mission foreordained by ancient ancestors. The only choice I had in the matter was to succeed or fail; alternatives, such as creating a destiny of my own choosing, would only result in failure. Okay, for the sake of being polite, I played along: so what is my mission? Answer: to help reconcile nature worship with monotheism, and thereby help heal the rift between Earth and Heaven that opened with Judaism, and widened with Christianity and then Islam. Sounds good ... can I mix in a little science as well? Netsere nodded, then added that I was to take up the work of an ancestor of, she explained delicately, low reputation. I had an uncle who was a bookie, but I knew that wasn't who she was talking about.

How she knew I was descended from Abu Jahal, my ancient forebear of some fifty-five generations back, I will never understand. Abu Jahal, whose name means "Father of Ignorance," is one of the greatest villains in the history of Islam. Three times my evil ancestor tried to kill the Prophet Muhammad, peace and honor be upon Him. Thanks to Allah for choosing Abu Jahal's nephew, Khalid ibn al-Walid, also my ancestor, to reject his treacherous uncle and instead lead Muhammad's army to the early victories that established Islam as one of the world's great religions.

It's one thing to have dubious heritage, quite another to be condemned to emulate it. What Netsere saw as valuable was the fact that Abu Jahal worshipped his ancestors. In fact, his first attempt at murdering Muhammad came when the Prophet declared that anyone who did not worship Allah would burn in Hell, including all ancestors who died before having an opportunity to learn about Allah. (This harsh doctrine is echoed in many other religions, including Christianity, which generally holds that entrance into heaven is barred to all who do not know Jesus Christ, including those who died before His coming to this world.)

Abu Jahal worshipped the Earth to which his ancestors had returned, and also the Moon and the stars that look down upon them—all of which the shamans such as Netsere see as pretty sensible spiritual practice. Less sensible, of course, was my ancestor's diehard advocacy of the infamous Satanic Verses, which would have heretically inserted praise for three goddesses, Al Lat, Al Uzza, and Manat, also known as the daughters of the Moon, into the Quran. Muhammad roundly rejected the pagan heresy in the name of Islam, a rigorously monotheistic religion. This prompted Abu Jahal's second attempt at assassination, causing Muhammad and his posse to flee from Mecca to Medina.

No way am I spending the rest of my time on this Earth living underground and wearing funny disguises the way Salman Rushdie, author of The Satanic Verses, did while under the fatwa, or death decree, from the Islamic mullahs. This I made perfectly clear to Netsere, a compassionate woman who kindly intimated that maybe I could, well, go light on the whole mission thing. Big-mouth me, I had to blurt out that my first book, *Gaia: The Growth of an Idea*, had been partly about nature worship. The die was cast, so she gave me an extra blessing and said something uplifting but decidedly unreassuring about the limits of this life.

Yana, my clever interpreter, and I almost joined our ancestors several days later. We were traveling from Irkutsk to Olkhon island in the southern part of Lake Baikal. Olkhon is Mecca for the shamans of the region, because it is magically sunny, weirdly energetic, and inexplicably warm. How Boris, our driver for the day, managed to avoid slamming into the black and white cow that popped out of the dense fog on the wet, slippery road, I have no idea. A few hours later, out on the shining island, an out-of-control open-topped jeep swerved to just miss hitting us head-on, then flipped over onto its occupants, who shrieked and shrieked and then wandered around in shock. An eagle, also known as a "tsar bird," soared up above us. In Siberian lore, the god who originally owned Olkhon turned his son into an eagle to investigate a rising religion, but then refused to turn his eagle-son back to human because he had eaten dead flesh. However, the eagle-son still got to inherit the island. Maybe that was him, taking in the show, I thought, flashing back to the black and white birds circling in the crystal blue sky the only time I ever visited my father's grave.

Boris consoled me that this sort of craziness does tend to happen when pursuing shamans. Those easiest to get to are usually the least able to speak authoritatively; some are out-and-out phonies. "Just because he has a drum doesn't mean he is a shaman," is how one local saying goes. Those hardest to get to are the ones you want to meet. In fact, identifying the most desirable shamans normally entails reverse psychology—find out who will not meet with you, then plug away. Feigning affliction of one sort or another would probably get you in the door, but then either quickly booted out upon being discovered as a phony, or worse, having to force your (healthy) mind and body to accept the shaman's powerful cure.

AFTERMATH

The true shaman's reticence about opening up to information seekers is in part to protect the uninitiated from handling such information irresponsibly. They see the act of sharing their secrets as analogous to teaching a five-year-old how to operate a gas stove. The child may easily master the mechanics, but that doesn't mean he won't hurt himself or, for that matter, burn the whole house down. Because of the high value shamans place on their specialized knowledge, they naturally assume that when they do reveal a tidbit or two, it will be taken very seriously by the recipient. They also assume that said knowledge will spread widely and quickly, much as I would expect my five-year-old, who might or might not listen to whatever else I had to say, to focus with laserlike attention as I taught him to turn the knobs of the stove, use the pilot light, etc., and then turn around and share this knowledge with his classmates in kindergarten.

Invariably, there are gatekeepers along the way who evaluate the prospective interviewer's character and/or relieve him or her of a bottle of spirits. Once in the presence of a true master shaman, the interviewer must be prepared for further evaluation of one's character, and also of what might be thought of as one's spiritual résumé, the history and circumstances of what brought one there specifically. In this instance, my personal disclosure gained me the entrée necessary to interview several distinguished shamans on Olkhon island that day.

"Twenty-twelve has already passed," said the eldest of the three shamans, call him Omula, as we sat on the porch of his humble, dusty home. Omula's native language was Evenk, an indigenous Siberian language, and his Russian was not great, so I figured that Yana had lost something in the translation. But no, as always, she got it right the first time. Yana explained the shaman's belief that the future exists in several distinct states. There is an independent future that is largely free and clear of the present, and therefore very much subject to change. But there is also what might be considered a dependent future, dependent on the present, and therefore with little chance of changing. Omula saw 2012 as a dependent future, with us here in the present drafting along behind it. I responded that I found his vision very discouraging, and the shaman asked me why. Because no chance for change means that nothing can be done, I replied. I could see that he believed the opposite, that there is a certain freedom in knowing that it's all in God's hands.

Just because the event stream that will culminate in 2012 has already been determined, as the shaman believes, the year need not turn out badly. I asked Omula what kind of a year it would be.

"The time period you ask about has been unkind to those who value progress over keeping faith with their ancestors," was how the translation came out.

Again with the ancestors . . . And what's so wrong with progress? To my way of thinking, "progress" and "future" are pretty much synonyms, although perhaps that needs to be reexamined. I wanted to know more about the future, but Omula was far more interested in my ancestors. Though not Muslim in heritage or belief, Omula was quite familiar with the pre-Islamic history of Mecca, and after a series of rapid-fire questions, I was astonished to find that yet another central Asian shaman had intuited my checkered ancestry. Had Omula and Netsere somehow been in contact? Couldn't say, though this man knew things that I never shared with Netsere, such as the fact that my female ancestor Saada Shehab, a benevolent young woman from about nine generations ago, had, with her father's patronage, in the 1830s established Lebanon's first Western-style medical school. Upon tracing my heritage back through Saada to Abu Jahal, Omula bolted from his chair, ran up a flight of stairs to the balcony, and then refused to come down.

Omula's younger colleagues took over the conversation, and the best I could gather from the freewheeling, breathless translation is that dedicated shamans such as the elder are healers first, and futurists barely at all. Because of this intense interpersonal focus, when asked to give general statements, as, for example, their opinions about 2012, shamans respond in ways that tend to have less to do with the substance of the question than with the person doing the asking. Omula's interest in me had to do with my ancestors, hence his answers to my questions about 2012 focused on the past. In five years of researching 2012 and its meaning for our future, not once had I ever encountered anyone who told me that the whole thing came down to whether or not we continued to ignore people dead and buried a long time ago. Yet here in Siberia, this was the message I was getting everywhere I turned. Confusing and depressing, it made so little sense . . . What was I going to tell my readers? Okay, everybody, go to the cemeteries, pray your heads off, and 2012 will turn out just fine?

WHY SURVIVE?

Most of us associate survivalists with "back to the land," crusty folks with rifles, canned goods, and MREs (meals-ready-to-eat), the crackle of a shortwave radio, maybe a hostage stashed in the root cellar, a Bible and/or some porn. Hunkering down in the hinterlands does have a certain escapist appeal, but unless you have access to land and know how to live off it, this is not a practical plan. The fact of the matter is that more than half the world's population lives in cities and almost certainly will continue to do so even in the event of a massive emergency. This section, therefore, will examine a range of survival skills that may well prove helpful in 2012 and thereafter, whatever the locale. Some of the alternatives are pretty gritty. May they never come in handy.

The most important survival skill is will. How strong is your will to survive? How strong should it be? The working assumption is that most of us are 100 percent committed to drawing breath as long as we possibly can, with a few exceptions: those who suffer irretrievably from terrible physical or mental illness, those who are so far advanced in years that quality of life has largely slipped away, and those who believe so firmly in the glories of the afterlife that they are pretty much marking time here on their way to Heaven. Let's assume that you fall into none of these asterisked categories. There are still profound variations in the priority people give to staying alive.

Please excuse the indelicacy of the following line of inquiry, but it is germane to the worst-case scenarios for 2012. Had you been in Hiroshima or Nagasaki when the atomic bomb was dropped, would you have preferred to perish along with the multitudes, or to have survived amid the horror, given the likelihood that the psychological and emotional quality of your life, however long, were tainted with post-traumatic stress disorder (PTSD)? How about if your projected lifespan were limited because of cancer or other painful illness caused by the radioactive fallout? How strongly does your preference depend on the survival status of your loved ones? Me, I'd rather go up in smoke, except if it meant leaving my children.

Perhaps you are the hero type, willing to subordinate your survival instincts to the higher calling of serving others and God. Father Mychal Judge, a chaplain with the New York City Fire Department, gave last rites to a dead fireman and then rushed back into the lobby of the stricken north tower of the World Trade Center, where he was killed by falling debris. Judge knew there was mortal danger in going back inside the tower, but risked his life anyway, and lost it. Sometimes it seems as though heroes want to die. There was a wonderful M*A*S*H episode in which a U.S. army sergeant of Korean descent had repeatedly risked his life to save others in his platoon as they fought in the Korean War. He did so because he was a courageous and honorable man, and also because, at an unconscious level, getting himself killed was the only way of resolving the conflict between his duty to serve his country and his abhorrence at killing the enemy, who shared his ethnic heritage.

During World War II, my father, who served in the U.S. Army's 88th Infantry Division, known as the Blue Devils, saw his two best army buddies die in foxholes near him during a battle in Italy. From that point on, I have always believed, a part of him wanted to die, so long as he could do it "responsibly," meaning that no one could blame him for chickening out on life, or purposely abandoning his duty to his loved ones. He loved cars, and died in a car crash twenty years after coming home from the war. Reason I bring it all up is because it makes me wonder how many of us grieve so deeply within that we would see a coming 2012 cataclysm as an escape clause from life, as an opportunity to die without being stigmatized as a suicide.

At the other insanely murderous extreme from heroes are those for whom the will to live has somehow been trumped by the will to kill: suicide bombers, manipulated into sacrificing their lives in exchange for next-world reward, for the promise that they will long be honored as heroes, and, much more tangibly, for the guarantee that they will have provided for the desperately impoverished families they have left behind. Bill Maher lost his show on ABC shortly after the attacks of September 11, 2001, when he declared that the suicide bombers who struck our country were not cowards. Homicidal maniacs, yes, but not cowards.

In the unlikely, unthinkable event that our nation were invaded, could you become a human bomb, if it meant taking out a bunch of oppressors? Probably not. But would you, could you, kill someone to save your own life or that of a loved one? An intruder? An innocent friend? An intruder I could kill without remorse, no longer the tender soul I was back in my college days, when we actually argued into the night about whether or not a person had the right, even in self-defense, to kill another. Now it's a no-brainer. Murdering a friend would of course be much harder to bring myself to do. But all bets are off if, God forbid, society were to collapse to the point where kill-or-be-killed became the rule of the day, such as in William Styron's World War II Germany. In *Sophie's Choice*, the protagonist was forced to choose which of two children to abandon to the gas chamber.

AFTERMATH

With the threat of social breakdown deepening as we approach 2012, we could find ourselves in situations analogous to Sophie's, forced to make unconscionable choices in macabre situations. What Sophie needed, what we all need, is a protector, someone whose will to survive is surpassed only by his or her skill at doing so, someone who is willing to do everything he or she can to make sure you survive too. Someone who thrives on the pressure, who has had lots of success fending for himself or herself in life-or-death situations.

TAKE REFUGE IN YOUR OWN LIFE

think for a moment...who of the people you know would be the most likely to survive a major, prolonged catastrophe? To whom would you run? How far away does that person live and how would you get there in the event of an emergency? What might they want from you and are you able to provide it? Do they owe you anything and are they likely to honor their debts? Who might compete with you for their protection? By what means could you triumph in that competition? Should your protector become incapacitated or unwilling to help, who is your backup protector?

If your first choice for protector was either yourself or a close family member, you are either fortunate or deluded. Think again, and don't be so wishful. And if you're too macho to believe you or your family would never need anyone's help staying alive, get over that right now. And as for waiting till the last minute and hiring professional survivalists, bodyguards, or the like, don't count on it: you may not be able to afford them since their prices will shoot through the roof, and their terms, particularly for people they don't already know, may include personal subjugation, although that might be a price you are willing to pay.

My advice is to attach yourself to whoever can best protect you and your family. Offer money, sex, friendship, the ability to read and understand complicated manuals, whatever works. (Hint: knowing how to make survivalists laugh is always valuable, since they can tend toward the grim.)

In a protector, one looks not just for skills but for potential, for grace under pressure. Me, I've been thinking of looking up Little Michael, a kid I grew up with in Park Slope, Brooklyn. Michael was not a very good student, kind of didn't see the point. He was not the type to score big on IQ tests, you know, where they show you a bunch of shapes pointing in different directions and you have to say which one is the opposite of the one in the box. Nor was he much for puzzles, riddles, and mazes. Math he wasn't bad at, especially when there were dollar signs next to the numbers. Reading was important, like the sports page in The Daily News.

Michael and his older brother, Richie, came from what today would be classified as a dysfunctional family, mother hanging out most of the time at the Bluebird Tavern, father away in prison—a rogue cop, they said, but I think he was just a rogue. So the boys mostly lived with their aunt, except every now and then they drove her crazy and she kicked them out of the apartment. One day, when Michael was about ten, he accidentally broke his aunt's special commemorative china plate that, if memory still serves, she got from the 1964 New York World's Fair. The pieces skittered everywhere. The pressure was on to fix the plate, perfectly, and that's exactly what Michael did. The aunt didn't notice a thing, and he and his brother got to remain indoors.

Piecing together that puzzle of china fragments was its own little IQ test, fitting shapes together, making the whole emerge from the sum of the parts. Would the A students at St. Savior's Elementary School have done as well as Michael in fixing the broken plate? Not necessarily. Some might have folded under the pressure—getting kicked out on the street is just in a different league from getting a bad grade. I wrote about

Michael once before, in my book *Common Sense: Why It's No Longer Common*, citing him as an example of good common sense. But that's not quite right. What he actually had is "survival sense." There was no alternative: if he wanted to avoid getting kicked out on the street, he had to fix the plate. If, for example, crying, pleading, and/or big-whopper promising could have mollified his aunt, that's an option he might have pursued, but he and Richie had already sucked that tit dry.

Periodic evictions helped Michael and Richie develop some pretty effective survival skills over the years. If mooching off friends or their alcoholic mom didn't work, it was time to raise some funds. Taking deliveries for the corner grocery store and shoveling snow in the winter paid pretty well but were not nearly as reliable as the New York City subway system. Michael would squiggle his way through the turnstile and then gimp around the F train, cadging for coins. I caught his act a few times and his fake pain spasms were deftly understated, really convincing. Once, one of the kids on our block got some free tickets to a Yankee game, seats right behind the official scorer's box, and a bunch of us went. Michael didn't have a cent in his pocket when we started out, but he made so much on the subway ride over to the stadium that he bought us all peanuts at the game.

The big money came from Richie, skinny, tough, and courtly, kind of like Fonzie but without a motorcycle or good looks. Richie always managed to offer a variety of items for sale, cherry bombs from Chinatown, individually wrapped mint-flavored toothpicks that sit-down restaurants sometimes kept in shot glasses near the register, even condoms, which he managed to sell to the guys on the block for a buck apiece even though, believe me, none us had any need for them. His specialty, though, was climbing up fire escape ladders and diving headfirst into garbage cans, charging everybody a quarter to see the show. Naturally, he broke his arm a few times, but somehow he always talked his doctor into overwrapping the cast into the size of a club so he could pound the crap out of anyone who tried to prey upon him in his weakened state.

Richie's crumpled wad and fake ID were good enough for the desk clerk in charge of the rooms at the local YMCA, which worked out well because the building was just down the block from the Bluebird Tavern, so the boys could drop by and visit their mother. After a couple days, Richie and Michael would talk their way back into their aunt's and things, as they say, were back to abnormal.

Last I heard, Michael was doing well, married a college girl and moved upstate. Richie fell, jumped, or was pushed off a roof and died. But both of them would be the kind of guys I'd stick very close to when the shit comes down in 2012 or whenever. They had the knack of working with people, albeit often with ulterior motives, to come up with whatever they needed to get by. Sure, Michael was a fraud and Richie was light-fingered, but only when things got desperate and, okay, maybe sometimes when things weren't really all that desperate. But they watched each other's backs.

No offense to Michael, but if Richie were alive I'd look him up first. Richie was the kind of guy who, say there were a sudden blast of nuclear radiation from a reactor meltdown, dirty bomb, or straight-out nuclear attack, would see it all as an opportunity, but he'd need some guidance on how to proceed. That's where I would come in. I'd tell him about potassium iodide, the best readily available protection from thyroid cancer, a horrible, quick-hitting, often fatal disease that radioactive fallout causes to proliferate wildly. Within twenty-four hours max, Richie would find out where he could lay his hands on a couple cases, then I'd do everything in my power to front him the money, gold, ammo, or whatever currency proved readiest in the state of emergency. Richie would then go out and barter or sell the pills and bring home the bacon for himself and whoever he was teamed up with.

How to stay on Richie's team once he's got your dough? Maybe all he would really need is a safe, welcoming home to return to, something he never had growing up. But why take the chance? I'd also prey on his insecurities. For example, Richie had zero confidence in his book smarts, so the thing to do would be to point out whenever possible that the need to read exists, and that you can fill that need. Worry him, for example, that the potassium iodide he is selling might be counterfeit, and then pretend that in college chemistry class, way above Richie's head, you'd learned how to test for the presence of iodine. Next either quick-study the procedure or phony up some lab-looking stuff, and selectively verify the quality of his inventory, pocketing the ostensibly fraudulent pills for your own private use. The reason for doing all this is simple: Richie was a good guy, and his marketing effort would have been seriously impeded if he knew he were selling bogus goods to people counting on him to stay alive. Now, if he could get a widow's Mercedes in exchange for a few $50 bottles of potassium iodide pills, he would take the car in a heartbeat. But for a bottle of placebos, even a loaf of stale bread would, to him, be highway robbery.

And when Richie no longer serves a purpose?

You get the idea. In extreme emergency times, one has to make oneself useful to those who can help one survive. Contracts, alliances, working agreements within your own network—in the end that approach will save a lot more butts than heading for the hills.

HIDE OUT IN BEVERLY HILLS

Crossroads for global glitterati, Beverly Hills serves as a microcosm of unbelievably rich and privileged folks everywhere, the ones best protected financially should global civilization face-plant anytime soon.

And also the ones least well equipped psychologically, say, to sleep behind a dumpster.

As a municipality, Beverly Hills is extremely well organized, with excellent police, fire, and ambulance services that will no doubt act heroically should catastrophe strike. This suggests an alternative for those who find the "cabin in the woods" 2012 catastrophe escape plan inconveniently rural: take the money that you would have spent on a bungalow and shotguns and instead rent the least expensive apartment available in Beverly Hills (where I have lived since 2001), or in Scarsdale, Marin County, Baden-Baden—whichever highly organized and affluent community is most convenient to your own. Make sure to obtain a primary lease in your own name, not a sublet. (Studio apartments are currently available in Beverly Hills for less than $1,000 per month.) Store the original, signed copy of the lease in your emergency escape luggage and be prepared to brandish it, along with corroborating government-issued photo ID, at the police-car roadblocks that will undoubtedly be set up at all entrances to your chosen community when pandemonium descends. If you can demonstrate that you are a legal resident, you and yours will likely be admitted by the police, especially if you are wearing business or cocktail attire. Don't take no for an answer.

Before committing to this course of action, however, it is helpful to familiarize oneself with overprivileged communities, whose customs and mores may, in fact, cause one to head for the hills.

"Have you thought about what you and your family are going to do in the event of a natural disaster such as an earthquake or terrorist attack? Are you prepared? Do you and your family and your neighbors have a plan to be self-sufficient for 3–5 days in case emergency services are not able to reach you?" asked the flier mailed to my home. For all the preaching I had done about the possibility of disaster in 2012, the honest answer was that no, I was not prepared, so I accepted the invitation from

David L. Snowden, Beverly Hills chief of police, to visit his department's Emergency Operations Center and attend "Don't Be Late in 2008," a seminar on how to cope with major catastrophe.

The thirty or so residents from Quadrant T7 in the flats of Beverly Hills sipped thoughtfully on their organic coffee as the duty officer methodically worked his way down the list of disasters. In addition to earthquake and terrorist attack, plague was a contender, including SARS and avian flu, as was tsunami, though primarily as fallout from beach areas such as Santa Monica that would take hit the directly. Riots, such as followed the verdict acquitting Los Angeles police officers of brutalizing Rodney King, were touched on as a possibility. A plan was presented detailing how residents should organize themselves and communicate with the police and other emergency services if cell phones and landline telephones were disabled. Special emphasis was placed on what to do, and what not to do, if power lines were down. Lists of basic foods and supplies to have on hand were provided.

Finally, the discussion was thrown open for questions, and one man's hand shot up immediately.

"Someone spray-painted graffiti in the alley behind my house. Can't you police do something to stop that?"

A golden moment of magnificent absurdity... The topic is how to prepare for death and destruction on a massive scale and this yutz is complaining about graffiti, which, by the way, BHPD comes and removes for free. Despite the duty officer's best efforts, the graffiti issue grew into a lively debate that took up most of the discussion period, that is until Rex, the police dog, came out and did some tricks for us.

The only explanation was denial: if all we ever talk about is graffiti, the earthquake/plague/terrorist attack will never occur. Denial is the psychological equivalent of the cartoon character who runs off a cliff and doesn't fall until he looks down. It provides much solace, and even a certain spiritual uplift, as though the truth were a headwind to be bucked by the brave. The more people have to lose, it seems, the more likely they are to deny the existence of threats to the status quo. And I say more power to them! Why get your bowels in an uproar?

The fact that you are reading this book would indicate that you are not in denial over the apocalyptic possibilities associated with 2012. But then again, the fact that those thirty or forty people attended the "Don't Be Late in 2008" BHPD disaster-preparedness seminar should have indicated that they too were ready to look the grimmest potential outcomes squarely in the face. There seems to be an approach-avoidance syndrome regarding the contemplation of major catastrophe. People are intrigued by the possibility because it is so dramatic, and also for the very good reason that learning about it could save their lives. But it's easy to go "tilt" when thinking the unthinkable.

Catastrophe, schmatastrophe. Honestly, which hits home harder: (a) the wholesale collapse of the North American power grid, or (b) a big, red, juicy nose pimple sprouting on the night of your birthday party? If you choose answer "b," consider seriously my suggestion to secure residence and take refuge in Beverly Hills, Scarsdale, Baden-Baden, or Marin when the shit hits the fan in 2012. There's a good chance that the cots in the high school gymnasium will be hypoallergenic and extra comfy.

LEARN FROM LEBANON

The good news and the bad news is that the average American is pretty low on the survival skill scale, because we have for most of our history been blessed with remarkable social stability. With the exception of the 9/11 attacks, the last major battles fought on our soil took place a century and a half ago, in the Civil War, which also saw the greatest number of

American war dead in our history. Of natural catastrophes we have of course had plenty, of which the aftermath of Hurricane Katrina ranks right down there with the worst. We have suffered our share of plagues, the Spanish influenza of 1919 and the current AIDS epidemic being examples of the deadliest. Violent crime is classified by the World Health Organization as a chronic health threat, and from that we have suffered far more grievously than any comparable nation. But overall, we have avoided the kind of disorder that could descend upon us in 2012 or thereabouts, and to prepare for that eventuality, it seems wise to turn outside our borders to societies that have more practice at coping with chaos.

Lebanon, the country of my ancestry, ranks high in chaos, having gone through several civil wars, massive military invasions by Israel, and wave upon wave of desperate and heavily armed Palestinian refugees. To illustrate the difference, ever since 9/11, lots of New Yorkers keep an emergency bag packed with essentials in case they have to get out in a hurry. Lots of Beirutis not only keep a bag packed, but take it with them whenever they leave the house, in case they get stranded.

My cousin Amale Saad hid out in a Lebanese Maronite Catholic monastery for several years during that country's civil war, 1975–1990. She was admitted while many others were turned away, because for several years beforehand, she had helped out the monks with small gifts and services, as well as faithfully attended religious services there. Monasteries are well suited as refuges because they are timeless; the people who live there dress, eat, and toil much as their forebears did for centuries, often in those very same buildings. Not beholden to modernity, many monasteries can function without rudimentary conveniences such as electricity or even running water, making them far less vulnerable to the disruptions of the outside world. They gain strength from their ancient traditions of reverence and simplicity and therefore seem a much more reliable long-term alternative than secular relief shelters, which though perhaps more copiously equipped, have no codes of conduct or belief so vital to long-term survival.

There are approximately fourteen thousand monasteries and convents in the world, mostly Catholic, Orthodox, Buddhist, and Hindu, together housing several hundred thousand residents, with the capacity for several hundred thousand more, should the need arise. Like the Von Trapp family in *The Sound of Music*, refugees from conflict or natural catastrophe have often sought sanctuary in these cloisters, though the number of supplicants usually far exceeds the capacity to shelter them.

So how to move to the head of the survivors' line? The moral of Amale's story, as we approach the potentially catastrophic year of 2012, is to plan ahead: choose the facility in which you might wish to take refuge, and then cultivate a relationship with its decision makers. That's the easy part. The hard part is, you will truly have to repent your sinful ways, cloistered clergy being remarkably sensitive to the stench of worldliness and spiritual corruption.

The Lebanese people are geniuses at surviving catastrophe. In 2007, I produced a small documentary called *Lebanon: Summer 2006*, examining that society in the aftermath of the war fought on its soil between Israel and Hezbollah, the Iranian-backed "party of God" that both serves and controls much of south Lebanon. The idea for the film came from Cedric Troadec, a Frenchman living in Los Angeles, who had gone to Beirut in July 2006 to attend his sister's wedding. While he was there, the Israeli invasion commenced, and he had to smuggle his sister, a new bride and seven months' pregnant, out of the country by boat to Cyprus and then on to a friend's apartment in Paris. When Cedric returned home, he was shocked at how biased and ignorant the United States news coverage of the Lebanon war was, so even though he had never made a film before, he resolved to do so to help set the record straight. Mutual friends introduced us, and in the autumn of 2006, I agreed to

Take Refuge in Your Own Life

assist him financially and creatively. At the time, it never occurred to me that the project would serve as a broader template for how people survive when the world around them crumbles.

"It was one hell of an experience," said Tima Khalil, a news producer in Beirut.

Being a people accustomed to war, at the first sign of major conflict, they sprang into routines. They stocked up on food at the supermarkets, and also on books, reasoning that even if the telephones, Internet, and electricity went out, they could still read by sunlight during the day. Pregnant women near term rushed to have caesarean sections performed before the hospitals were closed or bombed out. Those who could manage it got out of the country, many of them taking the same Cyprus-to-Paris route as Cedric.

Much of the aftermath of catastrophe, whether man-made or natural, is psychological, according to the survivors interviewed for the film. Loud noises such as the whistle and explosion of bombs, or the grinding of drones, rattled around in their heads for months, now years. Anger was a natural response, but hatred, most seemed aware, was to be resisted, primarily because of the injury it inflicts on those doing the hating. "He who hates tastes the poison first," the old adage goes. But some poisons are addictive . . . Paranoia and revulsion after Israeli warplanes dropped leaflets encouraging civilians to flee their villages, and then bombed the escape routes as they fled. Confusion about how to explain it all to the children who survived. Confusion about their own sick nostalgia for the war in the months after the battles ceased. Confusion.

"The consequences last for years," said Nayla Hachem, former director of the Lebanese Red Cross.

Little things grew way out of proportion, because sometimes the big things were just way too big to comprehend. One young man was crushed, couldn't get over it, that none of his friends called him on his birthday. Another man found that he could not bear the idea of being separated from his trumpet, so he carried it wherever he went.

The war brought a tremendous surge in creativity. Everyone became a writer, actor, painter, or singer; it was impossible not to express. Generosity bloomed, as one-quarter of the nation's population became refugees, over half of whom were taken by strangers into their homes. Sarcastic imitations abounded, frequently of Condoleezza Rice, roundly considered by the Lebanese people to have been a treacherous liar during and after the war for her complicity in supplying the invading army with cluster bombs, the worst sort of antipersonnel weapons that mostly end up maiming and killing scavengers, particularly children, all the while publicly declaring her compassion for innocent victims. A rare feeling of unity among the Catholic, Orthodox, Shia, Sunni, and Druse communities that make up Lebanon was forged by defiance, a victims-against-the-world sort of thing, and also by self-pity, a sense of being abandoned by the rest of the world.

The conceit of the film is a metaphor borrowed from the Lebanese writer Amin Maalouf, that Lebanon is a rosebush planted at the end of a vineyard row. Rosebushes are placed there because they are delicate plants, and thus early on show signs of infection that could go on to affect the grape vines. The implication is that Lebanon, being something of an amalgam of East and West, is, like the rosebush, an early indicator of troubles that could spread far beyond. Whether this analogy is prescient, or presumptuous, is hard to say. Suffice it to say here that should, God forbid, we in the United States ever face a wartime situation, may we all take a lesson from what happened in Lebanon during the civil war days of 1980, when the great singer Gloria Gaynor was brought in, driven right through a live combat zone to a lavish Beirut disco. The dancing, chanting, crying crowd made Gaynor sing her immortal hit, "I Will Survive," over and over again, into the night.

11

CONCLUSIONS AND RECOMMENDATIONS

a t its best, preparing for 2012 is kind of like preparing for graduation two or three years down the road. In high school and college, most of us did no more than whatever we had to do to make it to graduation, perhaps with a half-baked awareness that life would change pretty significantly afterward and that some allowances ought be made for that fact. But a few go-getters (I not among them) were always thinking a few steps ahead, about the terrific job for which they could make themselves the perfect candidate, about the business that they were going to start and then sell for big bucks before they hit thirty, about the voyage around the world that someone else would sponsor, about the marriage/kids/fabulous home package that would set them up for the rest of their life.

Frankly, I always looked down on these people, for their calculating nature, for their lack of ironic perspective on the absurdity of life, and mostly because I had neither the foresight nor the drive to be that way myself. Even today, having spent the last five years writing two books about 2012, I am not as prepared as I'd like to be for its dire eventualities, much less its rapturous possibilities. The following to-do list, therefore, is as much a note to self as to the reader.

MAKE A FAMILY PLAN

A friend, call her Kate, recently informed me that her family has made a plan to meet in case of major emergency, such as the failure of the electrical power grid. Regardless of whether or not the family members are able to communicate with each other prior to or during the emergency, their plan is for everyone to meet at the rambling old summerhouse Kate's parents own in the Appalachians. The house has a decent vegetable garden and sits on the shore of a crystal clear reservoir, thus solving the critical problem of gaining access to potable water. Plus, the reservoir is stocked with fish. The kitchen pantry has plenty of canned goods and the shed has got basic tools. There's even an old generator that maybe works. Kate's father is going to repair or replace it this summer, and lay in some extra fuel. There's even talk of everyone joining him up at the house for a week or two, mostly to get together and have some lakeside fun, but also to pitch in and help fix the old place up.

The problem is that Kate and her husband currently live in Los Angeles, about 2,000 miles away. Getting there by car in a time of emergency could be difficult, since gasoline might well be hard to obtain, particularly if the gas pumps lose their electricity. Air travel likewise would be hit-or-miss. So the challenge faced by this couple is to stay on top of the news, and to keep their lives flexible enough so that they can beat, even if just by a few days, any massive, frantic rush. For example, they would head for the hills at the first report of a space weather–induced blackout wherever it occurred, rather than waiting to see if the electricity also went down where they live. If signs of cataclysm start building toward a climax, Kate and her husband might proactively relocate to their family's home, though that would be a judgment call since it might mean losing their jobs, lifestyle, and current abode.

Conclusions and Recommendations

Few of us have capacious summer homes ideally situated and equipped to hunker down in during a major emergency. Then again, most of us are closer than 2,000 miles to wherever it is we will repair to be safe. So the particulars of Kate's situation may not be strictly germane. But we all have families, and a network of friends and associates, to whom we would turn in times of crisis, and who would turn to us.

Make a family emergency plan now. Decide where you would go, how you would get there, who would be welcome (and who wouldn't be). Is there someone, perhaps a protector/survivalist type whom, though not necessarily a member of the immediate family or inner circle of friends, you think it wise to include? What will that cost and how will you pay? Present your family emergency plan to prospective participants, discuss it, and agree on a course of action. It is helpful to introduce the idea slowly, first as simply a contingency plan for temporary emergencies lasting one week or less. In this instance, all that your family members would really have to commit to individually is having a bag packed and ready to go, plus pre-purchasing and stocking their share of food, water, and other household basics. Then gradually widen the scope of the discussion to longer-term and indefinite stays.

The next step is deciding where to go. Whether your plan is to gather where you currently live, head out for a neighbor's place, take refuge in a community shelter, or, like my friends, travel long distances, certain basic conditions have to be met. Your sanctuary must be under your control, or under the control of some person or entity, such as the Red Cross, whom you would trust with your life and that of your family. Basic personal safety from bodily assault must also be guaranteed as far as possible.

Some people feel absolutely the safest at home, no matter what the logistical shortcomings. Home-sanctuary folks can be a zealous lot, some preferring to risk death rather than be extracted. Their idea of preparing for emergency is preparing their home for emergency, hunkering down in order to repel the outside world, even if, say, a major hurricane is headed their way. Then there are the head-for-the-hills folks who, as noted earlier in this chapter, tend to romanticize the notion of escape but are often unrealistic about what it would take to scrabble up a life there, especially those who have no specific structure in which to take refuge, or skills to live off the land. Perhaps their thinking has been overly influenced by movies and other pop culture stories; as an antidote, they might reflect that most horror movies are in isolated settings. Being from New York City, my first instinct is "safety in numbers"; I'm more confident in my ability to deal with people than to fend for myself and my family in rural isolation. If I, life-or-death, had to negotiate with a crowd of strangers and make alliances among them, even if some spoke languages I didn't understand, that I could do. But if I had to take a shotgun, go out in the forest, and bring home the bacon, well, not so sure.

Comfort must always be outweighed by safety. Know in advance the extent to which you will be able to function if your refuge is disconnected from the electrical power grid and satellite communications. Is there a generator? Is there fuel for the generator? Being able to walk somewhere useful, such as the grocery store or emergency room, is preferable to having to rely on mass transit or a car in time of emergency.

Present the family emergency plan as a group project. Busy as we all are, many of us would nonetheless jump at the chance to actually do something constructive that would safeguard our future. For example, my friend's family sees preparing for a major emergency as an opportunity to spruce up their old summerhouse, a job that probably would have had to be done in a few years anyway. Though it might not be as fun or relaxing as the conventional vacations they might otherwise have taken, a good rule of thumb is that for every gallon of paint, a gallon of beer.

Some family members will dismiss the whole idea of a family emergency plan as unnecessary, preposterous. Rather than debating them directly, humor such folks as much as possible while coaxing from them the concession that IF some major emergency were to occur in 2012 or thereabouts, they would participate in the family plan. Do not be deterred if they persist in mocking the notion or even if they become angry and refuse to discuss any further the admittedly abhorrent notion. Fear and denial are eminently understandable reactions given the terrifying nature of the topic under discussion. Regardless of what they might say, these loved ones are included in the family's umbrella, whether they like it or not.

Writing up the plan and passing it around for signatures is probably unnecessary, except in complex agreements where different members agree to provide specific goods and services to the effort. Nonetheless, some sort of observance, such as lighting a candle, looking through family photo albums together, reading a sacred text, toasting to your future together, or simply sharing a moment of silence, will help the reality sink in.

SECURE THE BASICS

One major advantage of turning to shelters and other such community refuges is that the basic survival necessities of water and energy will be provided, presumably. Most of what follows, therefore, pertains to the needs of private refuges.

Procuring a supply of fresh water is of paramount importance in the midst of an emergency. Assuming no windfall such as the one enjoyed by Kate and her family with their reservoir-side summer retreat, the task at hand is twofold: to stockpile enough potable water to survive the initial chaos of emergency dislocation, and then to set up a longer-term system that provides a flow of water that is potable or that can be conveniently raised to potable levels. Water from deep wells is often drinkable, although relying on electrical pumps to transport it to the surface is risky, given the threats to the power grid. A backup hand-operated pump is a good idea. Cisterns, whether above ground or just below, have the opposite set of problems: access to the water is easy, but the water therein is often seriously polluted. Refuges near major highways or industrial areas should avoid dependence on cisterns.

Bacteria and other waterborne pathogens that come from sewage spills and other unsanitary events can largely be neutralized by the addition of iodine and/or chlorine water treatment compounds. Urine can also be recycled, treated, and reused as a last resort. It is interesting to note that some people in India, including former Prime Minister Morarji Desai, follow the practice of drinking their urine, with apparently no ill health effects. The surest way to kill waterborne pathogens is to boil the water for at least a minute. This, of course, requires an energy supply, much preferably one independent of the electrical power grid. Portable solar water heaters are generally incapable of raising water temperatures higher than 155 degrees F, so, at this stage of technology, larger, fixed-installation solar water heaters are about the only practical nonelectric alternatives capable of boiling water safely.

Of course, solar water heaters provide none of the electricity that might be required for other purposes. Solar-powered electrical generators tend to be large (12 to 15 square feet), heavy (30 to 35 pounds), and low-powered (150 to 200 watts), and they require the installation of a power converter, all of which is to say that the system would have to be expertly installed in advance. Photovoltaic cells, which store electrical energy collected by solar systems, are also expensive, though a nice addition to solar-powered electrical systems since these cells can supply electricity when sunshine is not available. Wind-powered systems have similarly daunting size and set-up requirements. Neither of these "clean energy" alternatives is as convenient or efficient as conventional generators, which are remarkably versatile and efficient. However, they depend on a supply of gasoline, kerosene, or propane, which may be difficult to obtain in time of emergency.

Food is easy: canned over freeze-dried or any other form that might draw on your limited supply of water. The opposite of organic, preservatives are good; the longer food can last without refrigeration the better. Dried fruits, nuts, anything with nutritional value that can be consumed without preparation. Halloween candy, individually wrapped and so heavily chemically preserved that it can be kept from one year to the next, works well.

Even if standard cell phone communications are disrupted, it still might be possible to communicate via satellite phone. Unlike conventional mobile phones, which transmit to local cellular towers that in turn amplify, package, and transmit their signals to satellites, satellite phones skip the tower and beam straight to outer space. This is important because conventional mobile phone networks tend to overload and break down during disaster situations, just when they are needed most. The bottlenecks occur at the cellular towers closest to the disasters, which, of course, can also damage the towers, causing further disruptions of service. While satellite phones are not foolproof or entirely immune to congestion problems, they do make for a good backup, and may prove to be a lifeline during an emergency. Solar backpacks designed to charge satellite phones are available for about $200.

For the basic hand tools necessary to survive under adverse conditions, *Popular Mechanics* (www.popularmechanics.com) lists fifty items, along with tips on how to use and care for them. Wrenches, hammers, screwdriver sets, saws, lubrication oil—there are so many "survival essentials" that one would probably need to add an extra room or shed to store them all. But it's a good reference to winnow and choose a dozen or so. It is helpful to obtain an expanded first aid kit, known in some regions as an earthquake survival kit, and also what might be thought of as a domestic survival kit, sewing supplies, SPF, batteries, candles, personal paper products, nonelectric toys, games, balls, books, and puzzles. Basic cold, allergy, and fever medications, multivitamins and antibiotics, potassium iodide pills to protect against thyroid cancer in case of radioactive fallout, even if they have expired.

Weapons are a matter of personal choice. While it stands to reason that with the descent of social chaos one might require firearms to protect oneself and one's family, such firearms are dangerous in untrained hands. So either take a firearms safety course or pass on the weaponry. An unloaded rifle can make for a good, bluff.

How to pay for all this? Everyone participating in your family emergency plan should agree to establish gift registries at Sears, Walmart, and other establishments where the vital emergency supplies can be obtained inexpensively. Make an agreement that some or all of the gifts exchanged between now and the end of 2012 come from that registry. Others, not included under the family emergency plan, should also be encouraged to select any Christmas, birthday, Mother's Day, etc., gifts they might give you from those registries. Yes, receiving a solar-powered shortwave radio is not as fun or sexy as Victoria's Secret, but there will be plenty of time to huddle together later on.

Economizing is in and of itself good preparation for emergency. Pilates may be the best way to make your body sexy and strong, but unless you are planning to buy and drag those infernal machines to your hideout, or show up with them at the relief shelter, save the $50 a pop and instead get used to a regime of push-ups, sit-ups, stretches, yoga, brisk walking or running outdoors—anything that can be done without special equipment and in the safety of one's home and immediate neighborhood. Similar savings can be realized on other personal care expenditures, such as manicures, pedicures, hair styling, and other pleasant luxuries that nonetheless are costly and will have to be provided, however inexpertly, by family members in the extent of a prolonged emergency.

In the event of a prolonged emergency, any of the gear listed herein may prove far more valuable than its original purchase price. You therefore may be able to barter any surpluses you might have for that which you lack. So if, for example, in preparing for an emergency stay, you have a windfall, say, of AA batteries or, to use the example from earlier in this chapter, potassium iodide pills, consider keeping them all for their potential trade value down the (rocky) road. Save whatever best suits your family's particular needs, then go to market with the rest.

DO GOD A FAVOR

Pray, meditate, channel past lives, implore extraterrestrial intelligences, propitiate ancestors, make burnt offerings. Unless you are into void and oblivion, do anything and everything to prepare yourself for a happy transition to whatever dimension of existence might come next.

If out of practice spiritually, ease back in as you might with any other personal relationship. Tell God a funny joke, inquire about His/Her day, offer to, well, just listen. Sure, such anthropomorphizing does a grave injustice to God's magnificence, but our humanity is pretty much all we mortals have, and I wonder how often we offer up even that pittance. Once I was trying to jostle this huge credenza away from the wall, and my scruffy little poodle, Max, kind of ran into it a few times, to help. Had no effect on the credenza, but I appreciated the gesture.

AFTERMATH

In the midst of a crisis, God just might look favorably upon those who, instead of praying for mercy or salvation, gave thanks, asked forgiveness, or otherwise distinguished themselves from the wailing, beseeching crowd. Not that one should strive to curry favor with God, exactly, just that good manners are always appreciated. Think of it from the Almighty's point of view: millions of souls crying for help, most of whom He/She hasn't heard much from until the chaos hit. What better moment to step up and take responsibility for whatever sins one has committed, and whatever role, no matter how small, one has had in the emergency that has befallen society. Climate catastrophes and greed-induced economic meltdowns are obvious examples of shared responsibility. Mea culpa, mea culpa, mea maxima culpa: there's no better time to own up to one's shortcomings than when those about you are just clamoring for help. Better yet, blow God's mind and in the midst of all the chaos thank Him/Her for your being alive, for all the blessings you've received in the past, whatever good stuff you can think of. Only caveat is that you've got to mean what you pray, otherwise you're just playing God for brownie points, which will backfire, I'm sure. Of course, getting in good with God doesn't guarantee results in this world or any other world, for that matter. Virtue, as they say, is its own reward. But at least it's a reward.

Unlike Roman Catholics, who must explicitly confess their sins to a priest in order to be forgiven, all we Episcopalians have to do is recite this pithy little prayer, which is of course adaptable for everyone's use: "Almighty God, Father of our Lord, Jesus Christ, maker of all things, judge of all men." Depending on your faith, you may, throughout this prayer, substitute the names of the deities whom/which you wish to implore.

"We acknowledge and bewail our manifold sins and wickedness, which we from time to time most grievously have committed, by thought, word and deed against Thy divine majesty, provoking most justly Thy wrath and indignation against us."

At this point in the prayer, one should specify the "manifold sins and wickedness" for which one seeks forgiveness. Just what constitutes sinfulness is far beyond the scope of our urgent purpose. Suffice it that there are two basic categories: (1) wrongful thoughts/words/deeds, and (2) lack of faith, aka despair.

"We do earnestly repent. We are heartily sorry for these our misdoings. The remembrance of them is grievous unto us. The burden of them is intolerable. Have mercy upon us, have mercy upon us, most merciful Father. For Thy son, our Lord Jesus Christ's sake, forgive us all that is past." Some theologies might also allow for one to get a head start and add a proactive, "And forgive us also for all the sins we may commit in the future."

"Grant us that we may ever hereafter serve and please Thee in the newness of life, in the honor and glory of Thy name. Through Jesus Christ, our Lord. Amen."

For help preparing your family emergency plan, and for other important guidelines relating to the 2012 aftermath, please go to lawrencejoseph.com. Please also check out www.ready.gov.

Aftermath Scenario: 2013

Nothing extraordinary happens on 12/21/12. Oh, there are any number of observances, events and personal epiphanies, and massive relief among those who had been caught up in the fear, but a nonevent is a nonevent, no matter how you slice it. It is kind of disappointing.

Mercifully, the postmortem palaver lasts only through the Christmas holidays; after New Year's Day, 1/1/13, it is back to the same old routine of trying to siphon electrical current, hacking into the satellites to make a phone call, wondering whether God or the Antichrist will appear, putting ten bucks into the betting pool of where the next natural catastrophe will strike or what industry the computer hackers will attack next. That sort of thing.

Weird events, such as the May reappearance of the hurricane cluster that had gathered off the West African coast the preceding autumn, are resolutely construed yet again as proof that civilization can take anything Nature can dish out. Ditto the spate of "quick-canes" suddenly popping up just off the southeastern coastline of the United States and striking within hours of their appearance. The pop philosophy "That which does not kill us makes us stronger" buoys countless spirits. Pat Benatar's 1980 chart-topper, "Hit Me with Your Best Shot," reemerges to become the unofficial anthem of the day.

Mark Twain's memorable observation that there are "lies, damned lies and statistics" is a help to those in denial over the millions of casualties now routinely being suffered. As though taking a cue from Holocaust-deniers, the Internet is awash with accusations and rumors that the numbers of dead and injured are being wildly inflated by powerful conspirators bent on sowing havoc and fear in order to seize control and impose martial law around the globe. A riptide countercurrent of bloggers insists on just the opposite, that casualty numbers are being depressed by those in power, to keep panic from sweeping the globe. Of course, not nearly as many people have access to the Internet as a just few years earlier, but those who do talk up what they discover there. Misinformation spreads by word of mouth, amplified and distorted to reflect whatever those doing the spreading believe.

Ballet Gulbenkian, a Portuguese modern dance company, picks up on this word-of-mouth phenomenon and choreographs it as a bodily expression of the old parlor game "Telephone," in which a person at one end of the room whispers a message into the ear of the next person, who then whispers it to the next person, and so on until the last person to receive the message says it aloud, comparing notes with the first speaker to see how distorted it has become in the process. In the modern dance version, the first dancer depicts a catastrophe with a series of explosive yet well-defined movements. Those movements then propagate down the line, with each successive dancer reenacting some of them while adding new ones, until the last dancer performs a whole new set of explosive movements while still somehow retaining the original sense of urgency and excitement. One night, while performing this dance in Lisbon, the last dancer in the "Telephone" piece steps down into the audience, and patrons begin doing their own version of the dance. From there, the Gulbenkian "Telephone" piece filters into Lisbon dance clubs, where it catches on as a popular dance step, with "messages" bodily transmitted from dancer to dancer around the floor. Soon "Telephone" makes its way into discos and parties around the Western world.

Is it a silly dance fad, just as the "Macarena" had been fifteen years earlier, or a physical manifestation for a new togetherness that goes deeper than words? Like toddlers first learning to use their limbs, practicing basic movements over and over again, "Telephone" dancers rediscover simple pleasure in their bodies' ability to move. The dance step helps folks pass the many hours they now spend in line waiting to get emergency supplies, medical care, and the few new jobs that are available. It also serves as an impromptu connector among the hordes of catastrophe refugees who do not share a common verbal language.

Time-honored concepts such as "meaning" and "intent" diminish in value in the post-2012 world. Partly, it's denial of the magnitude of the great social metamorphosis thrust upon the world, and partly it's a newfound exuberance of just being alive. Ancient conundrums, such as Jesus' declaration "Before Abraham was, I am," suddenly ring true. There was no time that is not, nor will there ever be. The ability to converse is now more important than the substance of the conversation. Singing, screeching, and making weird sounds all become socially acceptable forms of communication, for they all affirm the simple, joyous fact that communication is taking place. At the conversational level, free association becomes the rage. This time-honored diagnostic tool helps practitioners plumb their depths,

depths terribly roiled as a result of the catastrophic events. Plus, it's a fun and cost-free game to play as, say, folks wait for the power to come back on, gone, bye-bye, dearly departed, dearly beloved we are gathered here together today . . .

Optimism is no longer doctrinaire in the 2012 aftermath. The smiley face is dethroned. While the can-do spirit behind upbeat thinking is still admired, and the fact that a positive attitude does tend to beget positive results is still acknowledged, people no longer fear to admit sadness or fear. Unhappiness is no longer seen as the failure to be happy. For example, when asked how things are going, a real estate agent who hasn't made a sale in, say, thirteen months no longer feels compelled to say, "Never been better!" Instead she can feel free to swear like a truck driver just cut off by a Bentley with "TRSTFND" license plates.

There is a great sense of iconoclasm in the days that follow 2012. Museums come under attack, outrage from common folk who have been asked to swallow pretentious crap in the name of art. In one memorable episode, the Getty Malibu museum is stormed by a group of anarchists from Topanga Canyon, who work their way carefully through the premises until they find, and triumphantly destroy, a giant, ugly computer-chiseled sculpture of the head of Jim Dine, a narcissistic pop artist who had been commissioned by the museum to create art, and who came up with the bulbous, ruinously expensive self-portrait instead. Message from the vandals: never squander money that way again.

Morality . . . ha! "Right and wrong" is just not as important as "alive or dead." A tidal wave of despicably sinful acts serves as proof positive to religious fundamentalists that the time for the final Armageddon battle is nigh. But the zealots gain no traction. No one is in the mood to fight, particularly in the name of the Almighty, who has fallen into curious disrepute given all that humankind has suffered at the hands of His creation.

Political leaders around the world unite to deny their irrelevance, though their long-winded explanations are dismissed as so much global warming. Fears of a power grab or worse by China or Russia, nations less severely affected post-2012, are neither realized nor dissipated. In fact, the absence of aggressive political and military moves by these now semi-allied superpowers triggers deeper fears of some coordinated geopolitical thrust yet to come. For the first time in anyone's memory, the nations of the Middle East soften their chronic antagonism and even make some timid multilateral attempts at organizing global relief efforts. To many, a joint appearance on the world stage by the leaders of Iran and Israel serves as a shining example of hope and cooperation, but to the hard-core incredulous, it can only be a sign that kingdom is indeed about to come.

Enter the Antichrists, stage right, left, and center. The Revelation prophecy of Satan posing as Jesus has had an oracular effect, helping to prevent that of which it had warned. Everyone is mistrustful of leaders and so on guard against being duped that charismatic figures who might otherwise have risen to the fore are reflexively dismissed as phony or evil, whether or not they actually are. The cult figures and gurus that do pop up come from the oddest corners of the culture. Great whimsy greets the triumphal return of Jessica Smith, who in 1997 at the age of seven months appeared as the giggling "Baby Sun" at the start of each episode of *Teletubbies*, once an immensely popular television show for very young children. Smith's message that the Sun is our friend and need not be feared despite its recent "grumpiness" is taken seriously by no one yet somehow manages to provide laughter and comfort to many.

"And a little child shall lead them." Smith's reemergence coincides with a cult of youth focusing on the desperate hope that youngsters can somehow lead the world out of its post-2012 mess. Great expectations turn to "Indigo Children," a classification first articulated in the 1970s by parapsychologist Nancy Ann Tappe to the effect that, taken as a whole, children born since the Second World War represent a leap in the evolution of humanity's psychic and spiritual powers. Gifted misfits, Indigos are accounted to be disproportionately creative, troubled, brilliant, and rebellious, not unlike, some note, the psychographic profile for left-handed individuals. With each passing decade, Indigo numbers and power are purported to have grown in order to achieve their preordained mission of leading civilization to the next, enlightened age, which, according to the Mayan calendar at any rate, has just arrived. Though painfully short on scientific evidence, the Indigo hypothesis intuitively satisfies the perplexing question of why so many contemporary youngsters seem so weirdly prescient and wise beyond their years. It also jibes well with the Mayan prophecy that the souls of all who ever lived will return to physical existence in 2012—perhaps some of them in the form of these Indigo kids.

The color indigo becomes a political statement of defiance, revolution, and awareness. An Indigo tidal wave engulfs popular culture, led by a hipper-than-thou glitterati whose rejectionist cant is reminiscent of the "Don't trust anyone over thirty" sentiment of the late-1960s. Speechifiers extol Indigo Children as leaders, prophets, and gods, pointing to the indisputable fact that this new generation is indeed more knowledgeable about, more "into," the 2012 prophecies than their parents. Indeed, millions of high school and college students, Indigos and otherwise, had for the preceding few years read, watched, heard, and blogged about 2012, often while enduring the derision of their elders. A dangerous situation, where youngsters prove more knowledgeable than their elders about life-and-death matters.

The kids, predictably, get drunk with power and fat with praise. Remember your first hangover? Sober minds remind those hoping to be saved by the Indigos of what they always knew: there is no holiday from common sense, even during an apocalypse. Turning to children for inspiration is a wonderful idea, for guidance, less so, and for leadership, pathetic. Regardless of the color of their spiritual auras, children act out horribly when their parents abjure their responsibility to provide secure boundaries. And in the post-2012 chaos, that's just what all too many adults end up doing in their desperation for change.

Despite the generational debacle, the Indigos somehow manage to import the childlike wisdom that is their precious legacy. Indigo Children do indeed have a greater tendency than other children and than most adults to be psychic, telepathic, and clairvoyant, skills that come in handy in a world no longer able to depend so heavily on conventional electronic communications. While even the most gifted Indigos cannot, say, make a call without using a telephone, their sense of organic connectedness and shared thoughts preclude the necessity of transmitting and receiving gigabyte upon gigabyte of information, so much of which is redundant and unnecessary. Thanks to the Indigos, historians one day look back to 12/21/12 as the dawn of the Post-Information Age, or as the Maya would put it, the Ethereal Age.

What's coming next? People are desperate to know. A motley assortment of 2012 "prophets," claiming to have seen it all, come rushing in to control the future agenda. Sometimes positioned as "Indigo Adults," this "I told you so" contingent, particularly those who prove their street cred with (pre-2012) films, books, and websites predicting catastrophe, cannot be denied their moment in the Sun. After all, the mainstream intelligentsia for the most part flubbed 2012 big-time. Claiming that 2012 is only the start of a much greater transformative process, the 2012 prophets make a power grab for the global agenda. But like entrepreneurs whose start-up skills become obsolete when the organizations they founded reach the stage where more sophisticated management skills are needed, few of those who had foreseen the great changes coming in 2012 are now able to help much in shaping the aftermath. Past results are no guarantee of future performance, as they say. The 2012 prophets who assert radical change frighten off those who crave a return to normalcy. Conversely, those who counsel that civilization has only been suspended, not expelled for good, seem too soft, and inadequate to the Herculean task ahead. Preachers of prayer, thanks, and love are doomed to relearn the shopworn lesson that, sage as such counsel might be, people are not moved thus by words nearly as much as by heroic and loving deeds.

Not all 2012 contestants get bounced from the aftermath pageant: one guy who confesses how totally shocked he is that what he had predicted actually came true ends up carving out a niche for himself as the world's first "apocamedian."

APOCAMEDIAN

(Enters stage carrying a long metal rod with numerous batteries attached to top end. He also has two bottles of what appears to be water tied together and slung around his neck.)

Good evening, ladies and gentlemen. It's a pleasure to be here tonight at the Hilton Relocation Facility. I see you folks have electricity. I'll drink to that!

(slips something into his mouth, leans way back, then pours liquid from one of the bottles into his mouth. Foam bubbles and spurts up and out all over his face)

Volcanic eruptions. Remember all those predictions how the Yellowstone supervolcano was going to explode in 2012 and wipe us all out? Thing hasn't even burped. Now there are those who say we should stop drilling there, stop pricking fire holes into the lava balloon. Not me. I say keep stabbing that thing as hard as we can. Last thing we want is for Yellowstone to, you know, get comfy.

(leans back again into volcano position)

Oh, the herds of buffaloes, snorting and pounding. Thousands of hooves massaging my magma.

(repeats mouth-spurting routine)

Just call me Old Faithful.

(wipes face, drinks from water bottle)

Vinegar and baking soda, ladies and gentleman. Vinegar and baking soda for your entertainment pleasure.

(again wipes face, drinks from water bottle)

This?

(gestures with long metal pole)

It's my own personal lightning rod. My generator conked out about a year ago, so this is how I recharge my batteries.

(shakes batteries wired to the top of the pole)

If you think about it, the North American power grid is the largest lightning rod in the world. Very impressive achievement, very impressive. Only problem is that we forgot to let go.

(pantomimes electrocution, house lights dim)

(From darkness of stage we hear running steps and apocamedian's voice coming from different positions on stage.)

Over here. No, over here. Over here. No, over here.

(lights back up)

(Apocamedian has transformed lightning rod into a scythe, which, carried over the shoulder, makes him look like the Grim Reaper. From center stage he swings scythe at one imaginary evil after the other, felling them all. Finally satisfied, he throws the weapon down.)

Good night, ladies and gentleman. Better days are coming and our job is to be there to enjoy them when they do.

J ust before Christmas 2008, a friend, Ariane, called all excited to say that she had gotten engaged to her boyfriend, Alessandro. Not only that, she had a dream that I was the one who performed the wedding ceremony.

"So would you?" she asked.

"Would I what?"

"Marry us, you know, officiate at our wedding."

"But I'm not a licensed minister or judge or anything."

"That's okay, there are options."

Ariane explained that I could get legally licensed online by joining the Universal Life Church, but no way was I going to spend $75 or whatever to become some instant "Doctor of Divinity." That would jinx their wedding for sure. Otherwise, the county of Los Angeles has a program whereby, after paying $35, filling out a bunch of forms, taking a class, and getting sworn in, one is legally empowered to perform one specific wedding, of an aforenamed couple on the predetermined date. That sounded kind of neat and notarized. But shouldn't there be a law that no one who has been divorced is allowed to perform a wedding ceremony? Ariane knew my marriage had failed, so I decided not to bring it up if she didn't.

"Cool. Just let me know where I have to be when."

AFTERMATH

Sometimes when I can't stop thinking about something, it's because I am missing the point entirely, kind of like looking over and over again in the same place for a lost object, because that's where it's supposed to be, even though it clearly isn't there. After Ariane's phone call, I couldn't get my mind off this fancy silver and black Nehru jacket I had seen in a Sunday *New York Times Magazine* advertisement when I was around twelve. The ad was captioned "No black tie black tie." Maybe I got stuck on this memory because the jacket's round collar looked kind of like a priest's, or maybe it was because no matter how fancy the Nehru jacket was, it could never be the real black-tie thing. At the time, I was an altar boy at All Saints Episcopal Church in Park Slope, Brooklyn, and like most altar boys I sometimes wondered about entering the priesthood. Father Voelcker, the rector of our church, told me to deny the calling, and if it didn't go away after a year to come back and talk to him again. That settled that, or so I thought, until Ariane called four decades later.

Performing the interfaith marriage of Ariane, a secular Muslim, and Alessandro, a lapsed Roman Catholic, did seem in the spirit of rift-healing that Netsere, the Mongolian shaman I had met in Siberia, declared to be my mission in life, but how many bows can one take for going to a wedding? I decided to submit my application to the New Seminary in New York City, the oldest, largest, and most rigorous interfaith seminary in the United States, perhaps the world. Officially, interfaith ministers study and practice the world's major religious traditions, with an eye to finding commonalities and collaboration points among Vedic (Buddhist, Hindu, and related), Biblical/Quranic (Jewish, Christian, and Muslim), indigenous (including location-specific and nature-based practices around the world), ancestor-worship (Confucian and others), and even devout nonbelievers. Unofficially, interfaith ministers, less bounded by dogma and procedure than traditional clergy, make themselves available to guide and counsel those, and there are many, who believe in some form of God or Higher Power, but who are not sure what form that deity might take or who are not, because of their unorthodox lifestyle, welcome in the conventional religious settings they might otherwise choose.

The two-year New Seminary course of study will, if all goes as planned, lead to my ordination in June 2012. Graduation ceremonies now take place in Riverside Church, a great ecumenical congregation near Columbia University. Previously, the commencement was held in the Cathedral of St. John the Divine in New York City, where I was a parishioner for many years. Although nominally Episcopalian, the Cathedral, under the direction of Reverend James Parks Morton, was for decades a place where all manner of religionists, from lamas to rabbis to imams to shamans, were given the pulpit. I note with satisfaction the coincidence that John the Divine is the same John who lived in the cave in Patmos, Greece, and wrote the book of Revelation.

Becoming an interfaith minister would also address a question that I have been dodging for years, namely, What steps have I taken to prepare for 2012? I firmly believe that as 2012 approaches, and then with its aftermath, there will be unprecedented spiritual confusion and therefore an unprecedented need for comfort and internal peace. As religious minister, I could help those in distress, and also assist those who see 2012 the way the Maya did, as a rare opportunity for spiritual advancement. Unfortunately, your average interfaith minister is as poor as a church mouse without a church. Would that I could pay off mortgage and tuition bills with prayers, thanks, and love. We'll see how it all works out, with God as my guide and Mammon as my taskmaster.

CIRCLE OF MUD

All we know for sure right now is that we are alive, and that is truly a blessing. May the fear of losing that precious gift a few short years from now cause each one of us to treasure what we've got. To that end I propose a simple ceremony, wherein one person draws three circles of mud around another person's neck, in solemn observation of our connection to the Earth. Family members, friends, neighbors, strangers—anyone may anoint and anyone may be anointed. Doing so does not affect the celebrant's religious status, nor does it commit him or her to any set of beliefs.

The first circle of mud symbolizes a wedding ring, for the marriage of head to body, of sky to Earth, of birth to death.

The second circle symbolizes beheading of all thoughts that separate us from the goodness of life.

The third circle symbolizes eternity and the joy that comes with knowing that end follows beginning and beginning follows end, forever.

A simple act of togetherness in face of the gathering storm. And a reminder that whatever hardships we must face to get there, the future will always be ours.

ACKNOWLEDGMENTS

John Kappenman, an electrical engineer with great space weather expertise, enlightened me regarding the grave threat posed by solar blasts to the electrical power grid. The importance of this issue is impossible to overstate.

Andrew Stuart, my literary agent, operates at a high ethical and intellectual level, and once again got the job done very well indeed.

Roger Remy first got me thinking about the Sun's behavior.

Yana Kuznecova is a top-notch interpreter, translator, and cross-cultural navigator. The work she did locating and connecting to Russian, Mongolian, and Buryat shamans in Siberia was just terrific.

Brian McCourt, vidwiz extraordinaire, has a hard head, a good heart, and a kaleidoscope mind.

NOTES

Introduction

4   **"Impacts would be felt on interdependent"** National Academy of Sciences, *Severe Space Weather Events: Understanding Societal and Economic Impacts*, National Research Council, December 2008, p. 77.

5   **"It's about goddamn time!"** Rentilly, J., "Apocalypse Now?," *Mean Magazine*, meanmag.com, March/April 2007, p. 1.

8   **According to Carlos Barrios** Barrios, Carlos, *The Book of Destiny: Unlocking the Secrets of the Ancient Mayans and the Prophecy of 2012*, HarperOne, 2009, p. 136.

11   **"People who would never dream"** Yudkowsky, Eliezer, "Introduction," *Global Catastrophic Risks*, Bostrom, Nick, and Milan M. Cirkovic, Oxford University Press, 2008, p. 10.

19   **"This will happen in the last days"** Acts of the Apostles 2:17–21 (Revised Standard Version).

21   **"You see, my godfather"** Vanjaka, Zoran, and Jura Sever, *The Balkan Prophecy*, Vantage Press, 1998, p. 64.

22   **"When wildflowers lose their fragrance"** Ibid., p. 66.

Section I: What's in a Date?

25   **"It was the best of times"** Dickens, Charles, *A Tale of Two Cities*, 1859.

27   **"Fourteen hundred ninety-two is a year"** Reston, Jr., James, *Dogs of God:*

Columbus, the Inquisition, and the Defeat of the Moors, Doubleday, 2007, p. xix.

28 **"As if it weren't enough"** Barrios, Carlos, The Book of Destiny: Unlocking the Secrets of the Ancient Mayans and the Prophecy of 2012, HarperOne, 2009, p. 80.

Chapter 1: Here Comes the Sun

31 **"The electricity that attended this beautiful"** Original source is Philadelphia Evening Bulletin, as quoted in The New York Times, August 30, 1859.

31 **"Because of the interconnectedness"** National Academy of Sciences, Severe Space Weather Events: Understanding Societal and Economic Impacts, National Research Council, December 2008, p. 3.

32 **"Emergency services would be strained"** Ibid., 31.

33 **"The DOD is striving to increase"** Ibid., pp. 42, 48.

34 **"It was lamented that, in the eyes"** Ibid., p. 90.

35 **"Since the Space Age began in the 1950s"** Phillips, Tony, "How Low Can It Go? Sun Plunges into the Quietest Solar Minimum in a Century," www.nasa.gov, April 1, 2009.

36 **"We're just not used to this kind"** Ibid.

37 **In fact, according to Tony Phillips** Phillips, Tony, "Geomagnetic Megastorm," www.spaceweather.com, September 2, 2009.

39 **"This kind of influx"** Phillips, Tony, "A Giant Breach in Earth's Magnetic Field," a Science@NASA report, December 16, 2008.

39 **"The more particles, the more severe the storm"** Ibid.

39 **"It's the perfect sequence"** Ibid.

40   **"It's as if people knew there was a crack in the levee"** Phillips, Tony, "Sun Often 'Tears Out a Wall' in Earth's Solar Storm Shield," a Science@NASA report, December 16, 2008.

42   **"The [THEMIS] discovery overturns a long-standing"** Ibid.

Chapter 2: Down Goes the Power Grid

46   **"ACE has a prime view of the solar wind"** Christian, Eric R., and Andrew J. Davis, *Space Science Reviews*, vol. 86, 1998, 1.

49   **"The experience from contemporary space weather"** Kappenman, John, "The Vulnerability of the U.S. Electric Power Grid to Severe Space Weather Events, and Future Outlook," Prepared Testimony before U.S. House Subcommittee on Environment, Technology and Standards, Subcommittee Hearing, "What Is Space Weather and Who Should Forecast It?," October 30, 2003, p. 4.

53   **"Depending on the morphology"** Ibid., p. 7.

58   **"Political gridlock, broken markets"** Anderson, Chris, "Electrical Power Grid," Wired.com, April 2009, p. 3.

60   **"For example, the average number of hospitalized"** Krivelyova, Anna, and Cesare Robotti, "Playing the Field: Geomagnetic Storms and International Stock Markets," Atlanta Federal Reserve working paper, 2003, p. 4.

63   **"We have information, from multiple regions"** Gorman, Siobhan, "Electricity Grid in U.S. Penetrated by Spies," www.wsj.com, p. 2.

63   **"They [the Chinese] are all over the place"** Ibid., p. 2.

Chapter 3: Disconnected

69   **"For example, was a solar radio burst"** National Academy of Sciences, *Severe Space Weather Events: Understanding Societal and Economic Impacts*, National Research Council, December 2008, p. 41.

70   **"Radiation belt models are overly pessimistic"** Ibid., p. 61.

74   **"The consequences of war in space"** Myers, Steven Lee, "The Arms Race in Space May Be On," www.nytimes.com, March 9, 2008.

Chapter 4: Noah Returns

89   **"Our earth is degenerate these days"** Browne, Sylvia, with Lindsay Harrison, End of Days: Predictions and Prophecies About the End of the World, Dutton, 2008, p. 3.

90   **According to the Bible, "Wild animals"** Genesis 7:14–20 (Revised Standard Version).

90   **Second only to the story of Noah** Jowett, Benjamin, The Dialogues of Plato, vol. 3, Oxford University Press, 1871, 111.e.5 to 112.a.4.

94   **"During this time that the water was cooling"** Abbott, D. H., L. Burckle, P. Gerard-Little, W. Bruce Masse, and D. Breger, "Burckle Abyssal Impact Crater: Did This Impact Produce a Global Deluge?," in The Atlantis Hypothesis: Searching for a Lost Land, Papmarinopoulos, St. P., editor, Heliotopos Publishing, p. 182.

94   **"The archaeological record of this time period"** Ibid., p. 184.

95   **Moreover, those three metals** Blakeslee, Sandra, "Did an Asteroid Impact Cause an Ancient Tsunami?," www.nytimes.com, November 14, 2006, p. 2.

98   **Or as The New York Times observed** Ibid., p. 3.

Chapter 5: Sweltering in Siberia

113   **"a large-scale change in the climate system"** Clark, P. U., and A. J. Weaver, "Abrupt Climate Change," United States Climate Change Science Program, December 2008, p. 2.

116 **"The North Pole is melting"** Wilford, John Noble, "Ages-Old Icecap at North Pole Is Now Liquid, Scientists Find," www.nytimes.com, August 19, 2000, p. 1.

Chapter 6: Life Without Cucumbers

135 **They found that the stricken bees** Berenbaum, May, Reed M. Johnson, Jay D. Evans, and Gene E. Robinson, "Changes in Transcript Abundance Relating to Colony Collapse Disorder in Honey Bees," *Proceedings of the National Academy of Sciences*, August 24, 2009, vol. 106: 14790.

136 **IMD (imidacloprid)** Schacker, Michael, *A Spring Without Honeybees: How Colony Collapse Disorder Has Endangered Our Food Supply*, Lyon's Press, 2008, p. 2.

137 **"America's honeybees are sitting ducks"** Lockwood, Jeffrey A., *Six-Legged Soldiers: Using Insects as Weapons of War*, Oxford University Press, 2008, p. 250.

137 **"Each female [cucumber] blossom"** Ambrose, John T., Beekeeping Insect Pest Management, Note 7B, Cucumber Pollination, North Carolina State University, January 1995, p. 1.

138 **"Most important of all, is there a way to avoid"** Op. cit. Shacker, p. 4.

139 **"Are the honeybees trying"** Ibid., p. 5.

140 **"The ploughing of heathland and the draining"** Thomas, Jeremy, *Science*, March 2004, 303: 1879.

141 **"The results are appalling"** Ibid., 1880.

141 **"As the Earth warms"** Pounds, J. Alan, et al., "Widespread Amphibian Extinctions from Epidemic Disease Driven by Global Warming," *Nature*, January 12, 2006, 436: 162.

142   **"White-nose syndrome is a fungal infection"** Grant, Bob, "Deadly Bat Fungus Fingered," *The Scientist*, October 30, 2008, p. 1.

Chapter 7: Changing Climate Change

150   **"Global warming commandeers a disproportionate"** Bostrom, Nick, and Milan M. Cirkovic, *Global Catastrophic Risks*, Oxford University Press, 2008, p. 15.

152   **The economic potential is staggering** Kanter, James, www.nytimes.com, July 6, 2007, p. 1.

155   **"Pandemic disease is indisputably"** Op. cit. Bostrom and Cirkovic, p. 16.

155   **"If that happens, I will retire immediately"** ScienceInsider, May 4, 2009, p. 3.

157   **"IHR 2005"** Fidler, David P., Chinese Journal of International Law, Oxford University Press, September 5, 2005, p. 1.

159   **"We're being asked to buy an insurance policy"** Interview: Dr. S. Fred Singer, pbsonline, 2000, p. 4.

161   **"A review of the research literature"** Robinson, Arthur, Noah E. Robinson, and Willie Soon, "Environmental Effects of Increased Atmospheric Carbon Dioxide," *Journal of American Physicians and Surgeons*, Fall 2007, p. 79.

161   **"Claims of an epidemic of insect-borne diseases"** Ibid., p. 86.

162   **"I think one of the most pernicious aspects"** Durkin, Martin, producer, TV4, United Kingdom, October 8, 2007.

163   **Like his allies, he attacks** Dawidoff, Nicholas, *New York Times Magazine*, March 25, 2009, p. 35.

163 **James Hansen, head of NASA's Goddard** Ibid., p. 36.

163 **In his research monograph** Ibid., p. 57.

169 **"The size of the hydrate"** "Abrupt Climate Change," United States Climate Change Science Program, December 2008, p. 4.

Section III: Headed for the Hereafter

174 **Jesus lamented that the word of God** Matthew 13:18–23 (Revised Standard Version).

Chapter 8: Spelunking the Apocalypse

180 **"Happy is the man who reads"** Revelation 1:3 (Revised Standard Version).

180 **As it happens, the verse in Revelation** Ibid., 20:12.

182 **"Then over the earth, out of the smoke"** Ibid., 9:3–6.

183 **"In appearance the locusts were like horses"** Ibid., 9:7–11.

183 **Entomological warfare reached new levels** Lockwood, Jeffrey A., Six-Legged Soldiers: Using Insects as Weapons of War, Oxford University Press, 2008, p. 107.

186 **Wandered into a bookstore** Guillamont, A., The Gospel According to Thomas: Coptic Text Established and Translated, E. J. Brill, 1959.

186 **In Beyond Belief: The Secret Gospel of Thomas** Pagels, Elaine, Beyond Belief: The Secret Gospel of Thomas, Random House, 2003.

189 **"Heaven and earth will pass away"** Matthew 24:35–36 (Revised Standard Version).

192 **"Then I looked, and on Mount Zion"** Revelation 14:1–6 (Revised Standard Version).

Notes

193   **"The woman was clothed in purple"** Ibid., 17:3–6.

195   **In The Diary of Henry W. Ravenel** Ravenel, Henry, The Diary of Henry W. Ravenel, publisher unknown, 1876.

196   **"For my part, I give this warning"** Revelation 22:18–19 (Revised Standard Version).

Chapter 9: A Tale of Two Hemispheres

199   **"Shamanism is a special form of religion"** Kolyesnik, L. M., Shaman's Costumes, Irkutsk Museum of Regional Studies, APM, 2004, p. 6.

Section IV: Why Survive?

211   **In Sophie's Choice, the protagonist was** Styron, William, Sophie's Choice, Random House, 1979.

Chapter 10: Take Refuge in Your Own Life

214   **I wrote about Michael once before** Joseph, Lawrence E., Common Sense: Why It's No Longer Common, Addison-Wesley, 1994.

222   **In 2007, I produced a small documentary** Troadec, Cedric, director, Lebanon: Summer 2006, Sugar Bowl LLC Los Angeles, USA, and Lineamento, Aix-en-Provence, France, 2007.

223   **"It was one hell of an experience"** Ibid.

Chapter 11: Conclusions and Recommendations

231   **For the basic hand tools necessary to survive** "50 Tools Everyone Should Own (with Tips)," www.popularmechanics.com, May 2009.

234   **"Almighty God, Father of our Lord, Jesus Christ"** General Confession, Book of Common Prayer (1662), Church of England, p. 148.

Index

Index

Index

Index

Journalist and science consultant **LAWRENCE E. JOSEPH** has written about science, nature, and religion for the *New York Times*, Salon.com, and the Huffington Post and is the author of *Apocalypse 2012: An Investigation into Civilization's End* and three other books. His research on the science and mythology of Apocalypse 2012 during the past several years has led him to work with scientists, shamans, and philosophers on five continents. Joseph lives in Los Angeles with his two children.